Smart

Smart

A History of Intelligence

David Brydan

First published in the UK in 2026 by Footnote Press
An imprint of Bonnier Books UK
5th Floor, HYLO, 105 Bunhill Row,
London, EC1Y 8LZ

Copyright © David Brydan 2026

All rights reserved.

No part of this publication may be reproduced, stored or transmitted in any form or by any means, electronic, mechanical, photocopying or otherwise, without the prior written permission of the publisher.

The right of David Brydan to be identified as Author of this work has been asserted by him in accordance with the Copyright, Designs and Patents Act, 1988.

A CIP catalogue record for this book is available from the British Library.

Hardback ISBN: 978-1-80444-176-3
Trade paperback ISBN 978-1-80444-365-1

Also available as an ebook and an audiobook

1 3 5 7 9 10 8 6 4 2

Design and Typeset by IDSUK (Data Connection) Ltd
Printed and bound in Great Britain by CPI (UK) Ltd, Croydon CR0 4YY

Every reasonable effort has been made to trace copyright holders of material reproduced in this book, but if any have been inadvertently overlooked the publishers would be glad to hear from them.

The authorised representative in the EEA is
Bonnier Books UK (Ireland) Limited.
Registered office address:
Block B, The Crescent Building
Northwood, Santry
Dublin 9, D09 C6X8
Ireland
compliance@bonnierbooks.ie
www.bonnierbooks.co.uk

For Katie

Contents

Introduction		1
1	The Cognitive Elite	15
2	Intelligence Before IQ	62
3	The Birth of Intelligence Science	93
4	Mental Engineering	116
5	Mensa	160
6	Gifted and Talented	187
7	Building an Artificial Brain	225
8	Augmenting Intelligence	269
9	Resisting and Rethinking Intelligence	307
10	Why the History of Intelligence Matters	360
Acknowledgements		*369*
Notes		*371*
Index		*399*

Introduction

INTELLIGENCE – from the apparent genius of tech visionaries to the threat of omnipotent AI – is one of the most powerful and pervasive ideas in modern society. This is the story of how we came to think about it in the way that we do.

As soon as you start thinking about the idea of intelligence, I've found from my own experience, you quickly notice how often it pops up in everyday life. When I started the process of writing this book a few years ago I began to make notes on my phone whenever I read something where the idea came up.

In no particular order, here's a sample of some of the things I noted down:

> Former British Prime Minister Boris Johnson gave a speech in which he claimed human beings are deeply unequal in 'raw ability', pointing to differences in average IQ scores as evidence and arguing that this justified economic inequality. 'The harder you shake the pack,' he told his audience, 'the easier it will be for some cornflakes to get to the top.' (In this metaphor, as far as I understand it, clever people are the top cornflakes.)

Dominic Cummings, Johnson's former chief adviser, told an interviewer that the effectiveness of the British state could only be improved by 'very, very aggressively trying to get into position these very rare people who are times 100 or times 1,000 smarter than the norm'.

A new book reported that Barack Obama had initially welcomed the prospect of Donald Trump being chosen as the Republican candidate for the 2016 presidential election because Trump was 'nowhere near as clever' as his rival Ted Cruz.

Joe Biden challenged an Iowa voter to an IQ test during an argument at a presidential campaign event. Biden has form on this subject. When he first ran for the Democratic presidential nomination in 1987, he told a voter, 'I probably have a much higher IQ than you do', and then lied about how well he'd done at law school.

Donald Trump has said so many things about intelligence it is impossible to keep track of them all. On Kamala Harris, for example, he told reporters, 'I don't think she's a very bright person. I do feel that. I mean, I think that's right. I think I am a very bright person. A lot of people say that. I don't think she's a very bright person. And you know what? Our country needs a very smart person. And I don't think she's a very smart person.'

The mother of (some of) Elon Musk's children told Musk's biographer that 'he really wants smart people to have kids'. Musk talks regularly about his fears over declining birth rates.

In a meeting with Jeff Bezos, a senior Amazon executive said that the people leading strikes in Amazon warehouses were 'not smart or articulate'.

Ten Black people were murdered in a mass shooting at a supermarket in Buffalo, New York. The killer's manifesto stated that he was a white supremacist because 'I believe the White race is superior in the brain to all other races.'

The podcaster Joe Rogan founded a supplements company. It sells a product called Alpha Brain, which claims to boost cognitive functions like memory and mental speed. Meanwhile, the supermodel Bella Hadid launched a line of soft drinks which claim to 'enhance brain cognition'.

The release of ChatGPT-4 unleashed a tsunami of comments about artificial intelligence, including lots of claims about if and how quickly it was going to overtake human intelligence. One of my personal favourites was from Demis Hassabis, the Nobel Prize-winning CEO of Google DeepMind, who told Tony Blair in 2024 that current AI systems were 'not even at cat intelligence yet'.

A British educational campaign complained that a scheme supporting young people from underrepresented groups to apply to the University of Oxford 'discriminates against young people of ability and intelligence. Well-intentioned but misguided policymakers are placing diversity, equality and inclusion above academic ability because intelligence is not evenly and fairly distributed.'

What to make of all this? In some ways our apparent obsession with intelligence is surprising. Intelligence science and the idea of IQ – intelligence quotient – have become deeply contested over recent decades, mainly thanks to a series of controversies about race and IQ that began in the 1960s. People don't tend to embrace the language of IQ scores and intelligence tests in the way they did in the mid-20th century. They have become almost taboo.

But to see this as a rejection of the concept of intelligence as a whole, I would argue, is wrong. For a start, lots of people, as we can see from the examples given, still seem perfectly happy to talk about IQ. But more importantly, the way we think about intelligence in the modern world can't be reduced to the narrow notion of IQ. It involves a much broader set of ideas and assumptions which remains deeply embedded in today's society.

Let's start by considering how we encounter the idea of intelligence in our modern lives. This is something that can begin before we are even born. Our mothers, for example, might be told that what they choose to eat during pregnancy will affect the future cognitive development of their child. After we're born our parents will read about how playing with the right kind of toys, or listening to Mozart, or eating lots of oily fish, or any number of other things, will help their child to grow up to be smart.

Soon we'll begin going to school. For most people that will be the start of formal processes of testing, selection or streaming. It will also be the point when we might start to develop our own sense (through report cards, comments from teachers or comparison with our classmates) that there's some kind of scale of intellectual ability, and that we exist somewhere along it – that we're bright, or that we're slow. For some, this will be a big confidence boost. For others, it will be the beginning of lifelong feelings of inadequacy.

Eventually everyone will reach the stage of formal exams, where we will be publicly ranked and graded on our ability to learn. This is where our academic achievements will begin to harden into concrete life paths. In many parts of the world, success in these exams opens the doors to higher education, which increasingly means access to better jobs and higher incomes. In 2022, the OECD calculated that workers in developed countries with college or university degrees earned on average 50% more than those with only secondary (high school) education.[1] Even bigger rewards are on offer for those who perform so well in public exams that they're granted access to the elite tier of global universities, which in turn opens doors to the most highly paid and prestigious careers.

Failure, on the other hand, often limits people's career choices and their chances of future prosperity. Modern societies are increasingly divided between those who have higher education and those who don't. And in our meritocratic culture these divisions can take on a moralistic edge; those who succeed academically, many of us believe, succeed because they're smart, so they must deserve the rewards that come with their success.

What about beyond the realm of education? We'll certainly encounter the idea of intelligence in popular culture – through TV quiz shows, for example, or movies about tormented geniuses. We'll hear repeatedly that some of the richest and most successful people in society – the tech pioneers, CEOs or financial wizards – are rich and successful precisely because they're so smart. And particularly over recent years we'll have heard *a lot* about artificial intelligence, about the ways it's going to supplant human intelligence, and about what it will mean for our jobs, our societies or our very survival as a species.

Intelligence, then, is an idea that suffuses our world. We might not always notice it. Sometimes we might joke about it, or laugh at people who think they're cleverer than the rest of us. But it's an idea that's all around us, and one that we generally think matters.

For me, the trigger for thinking about the subject of intelligence was innocuous. I used to work in a job (a middle manager in an outsourcing company) where we talked about people's abilities in a certain way. When we did job interviews or performance reviews, we'd talk about what people were good at – person x is good at working with others, person y is good at managing their time, and so on. It's the kind of language that will be familiar to anyone who has a job with key performance indicators and annual appraisals. What we didn't do in that environment was talk about people's intelligence. You'd never really hear anyone say that Sarah in facilities was really smart, or that John in HR had a brilliant mind.

I then moved to another job – on a research project in a university – where people seemed to talk about ability in a different way. Now, perhaps contrary to popular stereotypes, it's not the case that people who work in universities are constantly walking around talking about how smart they are. Far from it. But there was a difference in the way we talked about people's mental qualities. There was a bit less 'good at', and a bit more of what seemed like code words for intelligence – colleague so-and-so is 'brilliant', student the-other is 'very bright'. Sometimes, people would refer to groups of students by their school grades – the need to recruit more 'A' students rather than 'B' students, with the implication that the former were fundamentally better in some way: better understanding, a greater capacity to learn, bigger brains.

These differences got me thinking about how we talk about the human mind and its powers. What do we think mental ability consists of? Why do we think about it in the way that we do? And why is it treated differently in different contexts? As a historian who has explored how our understanding of things like health, religion and international co-operation has developed over time, I knew it was impossible to answer these questions without turning to the past. To understand the ideas we carry around with us today we need to know where those ideas came from.

Like all ideas, intelligence has a history. That's not to say it doesn't have deep roots. Since the time of Plato, if not before, humans have been thinking about what it means to *think*, about whether some people are better at it than others, and about what that should mean for the way society is organised. But even until quite recently, intelligence wasn't understood in the same way it is today.

Imagine, for example, that instead of being citizens of the early 21st century we were instead living in the early 19th century. The idea of human mental capacity would still have existed, as would the idea that some people (or classes of people) possess more of it than others. But in the era before the development of psychology and psychometrics, IQ and intelligence science, intelligence would not have been understood in the same way it is today. In those societies still largely organised according to aristocratic principles, the idea that intellectual merit should determine someone's place in society wouldn't have been so compelling. When only some children attended school, mostly for just a few years of basic education, performance in academic tests clearly wouldn't have mattered to most people. And the average person's work or professional success would not yet be understood to rest primarily on their brainpower. If we went back further in time, these differences would be even more striking.

So how did we get from there to here? How and why did we go from a world where the idea of human intelligence was fairly vague and marginal to one today where it seems so important and pervasive? This book seeks to answer these questions by telling the story of how our ideas about human intelligence have developed since the turn of the 20th century.

Part of this story will encompass the history of IQ, intelligence science and eugenics. Some of this history has been explored before, by people like the evolutionary biologist Stephen J. Gould, who traced the history of scientific racism, craniology and IQ across the 19th and early 20th centuries in *The Mismeasure of Man*. This book, however, takes us much further. It is the first to place the science of IQ within the broader intellectual, political and cultural history of intelligence in the 20th and 21st centuries.[2] Alongside the work of psychologists and intelligence scientists, it draws on the archives of activists and international organisations; on diaries, newspapers, novels and interviews; and on the writings of computer scientists, educationalists, business thinkers and a whole host of others. These sources show how the story of intelligence extends far beyond the confines of IQ science, encompassing the history of schools and workplaces, of smart drugs and thinking machines, and of spelling bees and sperm banks. Only by telling this rich history of ideas and everyday life can we truly understand where our notions about intelligence come from.

This book also seeks to understand the story of intelligence from a more global perspective. Much of it focuses on Europe and North America, where intelligence science first developed and where the idea has had such an obvious social impact. But we'll also hear about the history of intelligence in places like China, the Soviet Union and Venezuela. It's not a comprehensively global history, however much it aspires to be. It rests partly on the research of others and, to be frank, we don't yet have enough of that research to tell a genuinely global history. What it does show is how intelligence became a matter of global concern during the 20th century, even though it wasn't (and indeed still isn't) an idea that was understood the same way in every part of the world.

Even with this expansive scope and global perspective, however, the history of intelligence is so rich and complex that only a part of it can be told here. The book is primarily interested in *high human* intelligence. We'll discuss artificial intelligence as it relates to the idea of human intelligence, but we won't spend much time on the intelligence of plants, animals or other beings. There are fascinating histories of *low* intelligence and the ways we've understood it, and there's a lot that could be said about *ordinary* human intelligence, although it's an idea that's often strangely overlooked. And lots of people have written about the idea of *genius* and its history across the ages. But what I'm mainly exploring here is how we understand the kind of high intelligence that we tend to label with words like 'clever', 'bright' or 'smart'. Above the average perhaps, but not completely out of the ordinary.

Beyond telling a history of the idea of intelligence, the book has another important task. It will show us how we came to see intelligence as something that *matters*. Part of the reason for this was that intelligence became deeply entangled with our ideas about progress. As our societies got bigger, messier and more complicated, people came to believe that those societies needed to be managed in new ways. No longer could we leave things to amateurs or aristocrats and assume the world would continue to progress as it had in the past. Instead, the ability, education and expertise of the most intelligent had to be harnessed to drive society forward. The new science of intelligence that emerged at the start of the 20th century promised to provide the ideas and tools to do just that.

But the main reason the idea of intelligence came to loom so large in modern societies was its relationship to the way we think about inequality. Intelligence, in its modern guise, came to be understood as a measure of difference – something that some people, or groups of people, possess more than others. It was an idea that was fundamentally hierarchical, placing some people at the top of the intelligence scale and others at the bottom.

This made it the perfect tool for the modern counter-revolution against equality. The 20th century saw the birth of a new global mass politics which, if not always egalitarian in the strictest sense, sought to upend existing hierarchies and promote new forms of equality and inclusion. In liberal democracies, the vote was extended beyond the narrow elites who had enjoyed it in the 19th century to include working people and women. Socialist movements and communist regimes loudly championed workers and peasants who for so long had been dismissed as ignorant by aristocratic and bourgeois elites. Anti-colonial movements fought to free their peoples from subjugation to European rule. Even fascism was built on a certain notion of equality within national and racial communities. All of these new political movements developed welfare and taxation systems which, to varying degrees, sought to equalise living standards and opportunities.

In the face of these trends, the modern idea of intelligence offered a golden opportunity for those who wanted to defend *inequality* and construct new hierarchies. People who worried that democracy would lead to anarchism and mob rule could argue that a new meritocratic elite was needed to keep popular passions in check. Defenders of colonialism and racial discrimination could argue that global inequality was caused by differences in intelligence levels between different races and regions. And the rich of all societies could defend themselves against demands for taxation or regulation by arguing that their wealth was a deserved reward for their exceptional talent and intellect.

These ideas about hierarchies of intelligence were powerful because they could be presented in such malleable and pernicious ways. They came cloaked in the language of fairness and objectivity, sold as ways of freeing societies from the shackles of aristocracy and nepotism. They could be adapted to changes in popular attitudes, linked to defective genes or to social disadvantage as the political wind dictated. Their defenders often avoided explicit mention of intelligence altogether, relying instead on the language of merit, skills or ability. Those labelled as intelligent had little interest in challenging the inequalities of the system. Those labelled as unintelligent often internalised that label, undermining their will to challenge the basis of their marginalisation. And where people did put up a challenge, their complaints could easily be dismissed as those of the ignorant or envious.

It's this malleability which explains why intelligence-based inequality continues to be seen as justified by so many people, even as other ideas about inequality have become politically unacceptable. Although ideas about intelligence have waxed and waned in popularity over the last century, and have continued to vary across different cultures and societies, they are still used as a way to justify modern patterns of inequality. They're a golden thread that runs throughout 20th-century history, and which continues to shape our world today.

What will we gain from understanding this history? It's tempting to think that it will give us some fundamental answers about the nature of human intelligence – whether it really exists, what it consists of, where it comes from, whether it's shaped by nature or nurture, and so on. But to be honest, it probably won't. Our ideas about intelligence have developed in particular ways for particularly historical reasons, but that doesn't tell us whether they're right or wrong, discovered or imagined.

But history can often help us most by throwing a splash of cold water over ideas and assumptions that might not, on examination, stand up to real scrutiny. What understanding this history *can* do, then, is give us some critical perspective, perhaps even a dose of healthy scepticism, towards the idea of intelligence. It might mean we prick up our ears when we hear someone or something described as 'smart', and make us wonder why that word is being used and what it's meant to convey. It might encourage us to reflect on the ways we think about talent, education and success in our societies today. It might help us to think through arguments about the nature of artificial intelligence and what it means for our future. Above all, it might offer us a platform to challenge those who seek to use ideas about intelligence to justify social and economic inequality.

It won't make us smarter, but it might help us to be a bit wiser.

1

The Cognitive Elite

'We hire very smart people and we pay them more than they think they are worth.'

Enron executive

LET'S START OUR STORY, not at the beginning, but at the end. Before we dive back into the history of intelligence, I want to think more about the way we have come to understand intelligence and its role in society over recent years. Because some time around the 1990s, a new set of attitudes and ideas towards intelligence began to emerge. We entered an age that appeared to venerate those with high mental abilities in ways that differed from what had gone before. It was an age when ideas about intelligence became even more bound up with the way we think about inequality and hierarchy than they had been in the past. It was an age that can be defined, in part, by the idea of the 'cognitive elite'.

The definitive book about the rise and fall of Enron is called *The Smartest Guys in the Room*.[1] The title is a comment on the way the men who built the company thought about themselves and their abilities. Over the course of the 1980s and 1990s, Enron transformed itself from a regional gas pipeline company into a hugely complex global financial behemoth. At the top of Enron's leadership group was a former McKinsey consultant by the name of Jeffrey Skilling. Skilling had a notoriously high regard for his own intelligence. At Enron, he built what he regarded as a genuine intellectual meritocracy, one founded on raw brainpower and creativity. Enron's value, he believed, came from its ability to develop complex new business models and financial innovations. The firm was using its intellectual firepower to upend an energy industry that had become mired in stagnation, conservatism and groupthink.

Skilling had little interest in recruiting people with relevant job or industry experience. Instead, he wanted to hire the smartest young people, usually the top-performing MBAs from business schools like Harvard and Wharton. He encouraged them to cycle rapidly through different roles in the company, rather than building up experience and expertise in particular sectors. And as the quote that kicks off this chapter suggests, he paid them lavishly. This was partly to discourage employees from leaving. But it also reflected Skilling's belief that the most talented people who took risks and made the most money for the firm deserved outsized rewards. It was a firm that saw its success as rooted in the talent and intelligence of its people.

The trouble with all this was that Enron's meteoric growth was based, in reality, not on talent and intelligence but on fraud and dodgy accounting. As the markets gradually came to realise this, Enron's financial situation began to unravel. It filed for bankruptcy in 2001, and Skilling and other senior executives were sentenced to jail. Either they weren't as smart as they had thought, or their faith in their own brilliance blinded them to the faults of the company they had built.

Despite Enron's fall, popular faith in the 'smartest guys in the room' and their importance to society continued to grow during the early years of the 21st century. A few years after Enron's bankruptcy, *The Economist* magazine published an article about the rise and rise of what it called the 'cognitive elite'. Those with the biggest brains, the article claimed, were now attracting ever-larger rewards because the most lucrative careers in fields like tech and finance required above-average mental skills. This trend was being exacerbated by new digital technologies, which meant that cleverness and cognitive skills were becoming more highly valued and rewarded. Because 'raw mental talents' are unequally distributed and the highly educated were increasingly tending to marry each other, inequality was surging and the wage premium for elite education was continuing to grow. The article raised some concerns about the effect of this growing inequality on social mobility and birth rates. But 'the clever', it concluded, 'shall inherit the earth'.[2]

The era of the cognitive elite is the one we still inhabit today. It's an era in which intelligence is regarded as the cardinal virtue, and a justification for ever-increasing inequality. The collapse of Enron might have revealed the hubris of its supposedly super-smart leadership, but the company's faith in the value of the highly intelligent was widely shared and deeply embedded across society. In recent decades the economic system has changed dramatically, becoming increasingly dominated by sectors like finance and tech. These sectors helped to feed a vertiginous cycle of economic boom and bust. Ever-greater rewards were offered to a highly educated elite, while the wealth and well-being of the rest of society stagnated. As these economic changes took root, the way we understood society and politics changed with them. At the heart of these changes was the idea that progress could only be driven by a rarefied social tier whose members had a particular set of advanced mental skills.

These ideas took root despite a certain public ambivalence towards the language of intelligence. This ambivalence was rooted in criticisms of intelligence science and the notion of IQ. Since the 1960s, a series of controversies over race and IQ had undermined public faith in intelligence science. These controversies were re-awakened in 1994 by the publication of a book called *The Bell Curve*. The fierce public debates about the book, which we'll discuss in more detail in Chapter 9, discredited the whole notion of IQ for many people. The emergence of ideas like multiple intelligences (MI), emotional intelligence (EQ) and neurodiversity helped to break down some of the more rigid ideas about human intelligence that had developed with the emergence of intelligence science in the early 20th century. The idea of IQ and the practice of intelligence testing didn't entirely disappear, but they played a much less prominent role in everyday life and popular culture than had been the case a few decades earlier.

The word itself was also being used less frequently. 'Intelligent' was increasingly being replaced in the English-speaking world with terms like 'smart', which boomed in usage from the late 1990s. In part this boom was due to the new talk about 'smart' machines in the digital age. But it also offered a way to continue to talk about intelligence without actually having to use a word that had accumulated so many negative connotations. Whereas in the past, the word 'smart' had suggested a kind of quick-witted, everyday ability to navigate the world (think of phrases like 'street smart'), now it increasingly came to be used as a euphemism for traditional ideas of intelligence defined by rationality, logic and pure brainpower.

But while the language people used to talk about intelligence changed, the ideas, by and large, remained the same. Even as the idea of IQ was becoming taboo, the unspoken, almost unsayable faith in intelligence was becoming even more apparent.

This faith in intelligence and the cognitive elite has, perhaps unsurprisingly, been held most firmly by elites themselves, a handy justification for their ever-increasing power and wealth. It has been evident across the political spectrum, from traditional liberals to the new breed of far-right populists. It has become embedded in the workplace, has reshaped the landscape of global higher education and launched the cult of the hyper-intelligent billionaire which has taken root in modern culture. It has become one of the defining principles of the 21st century.

The knowledge economy
The idea of a cognitive elite was nothing new. Indeed, we're going to see it re-emerge continually throughout this book, from Plato's notion of philosopher kings in ancient Greece to the debates about mass society in the 1920s, the creation of Mensa, and the expansion of gifted education after the Second World War. But it was turbocharged by new ideas about the economic system which began to emerge from the 1970s. Our renewed faith in the cognitive elite was inseparable from the development of the 'knowledge economy', and the ideas and industries associated with it.

In 1973, the American sociologist Daniel Bell published a book on the coming of what he called 'post-industrial society'.[3] Bell was reflecting on the disruption Western societies seemed to be facing in the 1970s, a period of major economic and social change, and was trying to predict what those societies would look like by the turn of the millennium. He argued that a fundamental change was afoot. Up until that point, developed countries had been *industrial* societies, organised around production and machinery. Now they were transforming into *post-industrial* societies. Rather than being driven by goods and manufacturing, future growth would come from services, from research and from knowledge. This would involve more and more people receiving higher education and working in scientific or technical jobs. By the year 2000, Bell predicted, the US and other developed countries would have become 'mass knowledge societies'.

Bell predicted that this post-industrial society would produce a social and economic hierarchy based on cognitive ability. At the top of that hierarchy would be a creative elite of scientists and managers, supported by a broader middle class of engineers, teachers and technicians. These skills-based elites would not just need to amass basic knowledge. They would have to master *theoretical* knowledge, and the abstract systems of symbols that were needed to put such knowledge to use. Scientific education would need to be opened up to the masses to make such a system work. But there would also need to be new systems to identify and cultivate the most talented, those with the intellectual skills needed to drive progress and guide this mass knowledge society forward.

These ideas, relatively new in the 1970s, had firmly taken hold by the 1990s. Bill Clinton's Secretary of Labor, Robert Reich, described a new economy dominated by a minority of 'symbolic analysts' – management consultants, software engineers, financial advisors and others who worked in networked, intellectually challenging roles.[4] This economy was a globalised one, rooted in new communication technologies which enabled these symbolic analysts to work across national borders. On the other side of the

Atlantic, politicians like Tony Blair were urging their subjects to embrace this new globalised knowledge economy. 'The key capability for people to survive and thrive,' Blair argued, 'is their capacity to learn.'[5]

These ideas about a post-industrial, globalised knowledge economy were accompanied by a new political vision of technocracy, or the rule of experts. This was something Daniel Bell had also predicted: as the knowledge society developed, as economic life became more complex and science more important, the old political ideologies would wither away. All that would be left were technical questions, the ideal policies needed to effectively manage and plan the new society. These could best be decided by skilled, highly educated experts. If politics was all about making smart decisions, it made sense to put the smartest people in charge.

Technocracy wasn't a new idea. Some of the utopian socialists of the 19th century had believed that technocratic rule could replace class conflict. If the rulers of society were genuine experts, everyone would be happy to obey them in the same way people follow the advice of their doctor. The term itself was coined at the start of the 20th century, inspired by the idea of modern society as a machine that needed skilled engineers to make it run efficiently. Figures on both left and right imagined a new, rational industrial society in which the technicians would take the place of politicians. These ideas soon fed into the practice of politics. Franklin D. Roosevelt assembled a 'Brain Trust' of experts to advise him during the New

Deal. John F. Kennedy brought the 'best and brightest' from academia and industry into his administration to run foreign policy (albeit with pretty disastrous consequences).

But the idea of technocracy reached its peak in the late 20th and early 21st century. The end of the Cold War reduced the ideological tension in global politics. Western politicians believed that they had won the Cold War partly because they were simply better at managing their societies and economies than their ideologically driven Soviet counterparts. As faith in the power of markets grew, politics seemed to be less a contest of ideologies or values, and more a question of who had the best policies to ensure those markets worked efficiently. Politicians like Bill Clinton and Barack Obama loved to describe policies as 'smart', and frequently used phrases like 'It's not just the right thing to do, it's the smart thing to do.' As the American philosopher Michael Sandel has argued, the politics of this era became little more than 'technocratic talk'.[6]

This technocratic vision of expert rule was reflected in a big change in the make-up of the political class. Under Clement Attlee's British government that was elected in 1945, 43 per cent of Labour MPs had left school before the age of 14.[7] By the end of the century almost every politician and member of parliament in every Western country had been to university, with many holding advanced degrees. British Prime Minister Gordon Brown invited experts from the business world into his cabinet and boasted of creating a 'Government of All the Talents'.[8] Twenty-first-century Italy twice turned to the technocratic

'Super Marios' to drive through austerity programmes and restore its credibility with the EU, first with the economist and European bureaucrat Mario Monti during the Eurozone crisis of 2011, and then Mario Draghi, the economist turned international banking expert, during the Covid pandemic of 2021. Smart politicians, so the thinking went, would lead to smart politics.

New ideas about economics and politics were reinforcing the belief that society needed to be led by the intelligent, by a class of cognitive elites. This renewed faith in the importance of the educated and intelligent spread beyond the world of politics. It also became embedded in the new industries that dominated advanced economies in the post-industrial age.

Take management consulting, for example. Today it is a multi-billion-dollar global industry. But it is also a fairly recent invention – 'the world's newest profession', in the words of one of its historians.[9] The industry emerged in the United States in the 1930s, part of the drive to turn management into a scientific profession rather than leaving it to amateurs in the form of company founders or owners. Early consultants referred to themselves as 'management engineers', spreading new ideas and techniques like Frederick Taylor's scientific management across corporate America.

But in order to survive, management consultants had to convince the leaders of corporate America that they had something useful to sell. Consultants had no product to offer. All they had was themselves. So they had to find a way to commodify the figure of the consultant, his (in those early years consultants were always men) personality, his knowledge and, above all, his intelligence.

We can see how important the idea of intelligence was to management consulting through the history of McKinsey. Since its foundation in the United States in 1926, McKinsey has grown into one of the most well-regarded and influential consultancies in the world. It has shaped the policies of governments. It led the drive to increase executive pay levels around the world. And it invented financial technologies like securities, which have driven economic growth and repeatedly plunged us into financial crisis.

McKinsey also has the dubious distinction of being the very first company in the world to try to recruit through Mensa, the society for people with high IQs whose history we will discuss in Chapter 5. McKinsey (or *The Firm*, as it likes to refer to itself) had opened its first overseas office in London in 1959. It had originally aimed to recruit all of its consultants from Oxford and Cambridge. But in 1963 it decided that Mensa, then little more than a small group of English eccentrics, represented another pool of potential consulting talent. Placing the first ever recruitment ad in Mensa's magazine, it promised members that McKinsey was an intellectual meritocracy, where 'progress depends solely on individual ability, effort, mental superiority, and performance'.[10]

This was very much in keeping with McKinsey's self-image as an organisation defined by its intelligence. McKinsey-ites saw themselves as the intellectuals of the business world. In its early decades, consultants were required to read 15 books a year from an approved list and submit book reports to partners. Internal memos described McKinsey employees as 'the best and the brightest'. The British politician and former McKinsey-ite William Hague described the company as a place where 'you are encouraged to believe that you belong to a special group of elite people'.[11] Its guiding principle was to hire the smartest people, give them interesting work to do and largely leave them to get on with it.

This was reinforced by McKinsey's commitment to recruiting only those with elite education. In the 1950s it forged a close relationship with Harvard Business School. Up until that point, business schools and MBAs had a mixed reputation within corporate America. But McKinsey helped to establish the idea of elite business education as a mark of intellectual superiority and business acumen. By the 1960s Harvard Business School had become McKinsey's primary recruiting ground, with a particular focus on the Baker Scholars, who graduated in the top 5 per cent of their class. Once, by the 1990s, this strategy had become widely copied by other elite consulting and financial firms, McKinsey pivoted instead to hiring science and engineering PhDs. As another McKinsey job ad put it, all that was needed to succeed in The Firm was 'outstanding mental equipment finely honed by a first-rate education'.[12]

A similar cult of intelligence was evident in the financial sector, another of the dominant industries in the modern knowledge economy. In the mid-20th century, the banking sector had appeared to be in decline. It was largely a middle-class profession, often derided for its bureaucratic nature and relaxed work ethic. In 1941, less than 2 per cent of Harvard Business School graduates chose to work on Wall Street.[13] But the liberalisation of banking from the 1980s sparked a boom in the size and dynamism of the financial sector. Suddenly, an industry that had prized stability and dependability began to recast itself as one of entrepreneurship and cutting-edge intellect. As one investment banker in the City put it, the finance industry boomed from the 1980s because it became meritocratic, attracting 'the biggest brains, the best intellectual capital'.[14]

These trends were exacerbated in the 1990s and early 2000s with the rise of technology and data in finance. Firms began developing their own proprietary algorithms to run trading portfolios. The complexity of markets massively increased, with new products like derivatives, swaps and securities requiring more detailed analysis and more accurate forecasting models. Competitive advantage increasingly lay in the predictive power of a company's financial models. Firms vied for the most highly rated maths and economics PhDs to help develop them. 'These days,' the *Economist* argued in the mid-1990s, 'the advantage lies in the brainpower and associated computing power that firms can mobilise.'[15]

The industry that became most closely associated with the idea of a new cognitive elite from the 1990s was tech. We will see later in this book how the history of Silicon Valley was entwined with the history of intelligence science.

Stanford University, at the heart of Silicon Valley, was central to the development of intelligence science in the US, and thus to spreading a 20th-century vision of intelligence as the key measure of individual worth. William Shockley, the inventor of the transistor and founder of the tech sector in Silicon Valley, was one of the leading proponents of intelligence-based scientific racism from the 1960s.

This legacy was evident in the way that the tech sector came to revere intelligence. Silicon Valley was built around the cult of individual effort and ability. It saw itself as meritocratic, indeed as the only truly meritocratic part of society. It was a place where old hierarchies and social structures supposedly had no purchase. In tech, so the story went, your age, race or gender was irrelevant (despite the fact that the industry was – and is – dominated by men). All that mattered was your brainpower and your willingness to work hard.

As such, industry leaders saw their key task as hiring the most intelligent workers. It was those workers who would develop the intellectual property on which the fortunes of their companies could be built. Mark Zuckerberg said that the most important thing he looked for in hiring people was 'raw intelligence'.[16] Steve Jobs argued that his success was built on hiring truly gifted people, people he referred to as 'A players' (rather than B or C players). 'For most things in life,' Jobs told one of his biographers, 'the range between best and average is 30 per cent or so. The best airplane flight, the best meal, they may be 30 per cent better than your average one.'[17] But the best people, he believed, could be 50 times better than the average. These were the A players he tried to build his teams around.

Nowhere was the value placed on high intelligence more evident than in the cult of the tech founder. Media reports and books on Silicon Valley since the 1990s have lauded the intellectual power of those who started and ran the leading tech firms. The venture capitalists at Y Combinator have claimed that they fund not great ideas or companies but 'smart founders'.[18] In a similar vein, Peter Thiel's Founders Fund says that its mission is to 'invest in smart people solving difficult problems'.[19] And many of those founders seemed to have had an equally high regard for their own brainpower. Paul Allen, the co-founder of Microsoft, said of the young Bill Gates that '[h]e was really smart. He was really competitive; he wanted to show you how smart he was.'[20] Gates worked closely with Steve Jobs during the early years of Apple, but their relationship was always fractious. 'Each one,' according to one Apple employee at the time, 'thought he was smarter than the other one'.[21]

The worship of intelligence in the tech sector was slightly more nuanced than it was in other industries. In a 1982 speech to young people, for example, Steve Jobs reflected on what it meant to be intelligent. He made little attempt to hide his self-regard on the subject ('You meet your friend and you think they're dumb and you're smarter than them,' he told his audience in one anecdote, presumably assuming this is an entirely normal thing for someone to think about their friends). But he rejected the idea that intelligence was just a matter of brainpower. It was also, he believed, about memory, about the ability to zoom out and see the bigger picture, and about making innovative connections. To do that, he believed, you had to open yourself up to different experiences. 'You might want to think about going to Paris and being a poet,' he told his audience, 'go to a Third World country . . . and see people, lepers with their hands falling off . . . Fall in love with two people at once . . . Walt Disney took LSD . . . and that's where the idea of *Fantasia* came from.'[22] Jobs believed in the power of a cognitive elite, but he wanted the members of that elite to seek out experiences that would help them to combine their raw brainpower with new and unusual perspectives.

Such attitudes were partly a reflection of Silicon Valley's roots in the California counter-culture of the 1960s and 1970s. The region had been one of the breeding grounds for new ideas about expanding human potential through new experiences, drugs or different ways of living, as will be discussed later in the book. They also reflected the symbolic importance of the college drop-out in the cult of the tech founder, a thread running from Jobs to Zuckerberg. The figure of the drop-out was always an ambiguous one. They could be seen either as an educational failure, or as someone who was *so smart* they didn't even need the elite education they'd gained access to.

Silicon Valley certainly valued elite education. Colleges like Stanford offered a reliable route for young people to secure jobs there. But the tech sector had a more ambivalent attitude towards the credentialism of elite colleges than industries like consultancy or finance. And tech leaders almost always combined their lionisation of high intelligence with the argument that it needed to be combined with other characteristics the sector valued highly, things like creativity, the willingness to adapt and to disrupt, and the ability to work punishingly hard. Even Bill Gates, who said that when he was young he wanted to build an 'IQ hierarchy' at his company where everyone would be managed by a boss who was smarter than them, came later to admit he was wrong and that 'IQ seems to come in different forms.'[23]

These slightly more nuanced celebrations of intelligence became embedded in the new ideas about cognitive elites which emerged from the 1990s. These elites were now rarely described through the language of IQ. Notions like MI, EQ and collective intelligence became embedded in the way mental abilities were discussed, even in those industries that most explicitly valued brainpower. Intelligence was still celebrated – even celebrated more than it had been in previous decades – but with a different language and in combination with a wider range of skills and characteristics.

All of this fed into a corporate culture in which 'talent' was elevated to a cult-like status. True to its origins and self-image, McKinsey was central to this elevation of talent in the corporate world. In the late 1990s a group of its leading consultants published a hugely influential series of reports on what it called 'The War for Talent'.[24] Talent management, the reports warned, needed to become the burning corporate priority. In the increasingly complex modern economy, a firm's ability to attract and retain superior talent was the key source of its competitive advantage. Corporate leaders needed to be personally involved in hiring. They needed to support and mentor the most talented people in their organisations. And they should ruthlessly weed out mediocre performers who risked blocking the professional trajectory of the talented elite.

Above all, the McKinsey consultants argued, companies needed to better reward the most talented. This meant ensuring their pay was considerably higher than that of averagely performing colleagues. But it also meant offering them a long-term route to wealth accumulation. These arguments helped bolster the trend of ever-higher executive pay, and led to development of the internal stock option schemes which have now become a ubiquitous part of the elite economy. These trends have helped turbocharge inequality over recent years.

Talent, for the McKinsey consultants, wasn't reducible to intelligence. The complexity of modern business life required global, multicultural thinking, technical skills and entrepreneurialism. Firms would have to decide exactly what kinds of talent they needed in order to thrive. But the language of intelligence was almost always a part of the way talent was described, even if companies said they wanted people who combined raw brainpower with other character traits. One corporate leader told McKinsey that he asked himself three questions about potential recruits: 'Are they really smart? Can they get things done under pressure and in unfamiliar circumstances? Are they nice?'[25] Intelligence was no longer the be-all and end-all of corporate recruitment, but it still very much mattered.

The poster child for this new corporate approach to talent, and the firm that was most frequently cited in the McKinsey talent reports, was Enron. Enron began life as a fairly run-of-the-mill gas pipeline business based in Houston. In the late 1980s, helped by the deregulation of the energy sector in the US, it began to transform itself into an energy trading company, moving gradually but inexorably from infrastructure to finance. From initial experiments with oil trading, it ultimately developed into a hugely complex trading business, dealing with derivatives, swaps, securities and the array of other products and innovations that emerged from the finance boom of the 1990s. By that point, Enron had become one of the most profitable, fastest-growing and admired companies in the world. It regularly topped charts for the best and most innovative firms.

As we saw at the start of the chapter, Enron's celebration of talent and intelligence was evident in the self-perception of its leaders and its approach to hiring. It was also embedded in an intensive, cut-throat performance review system. Twice a year, every Enron employee, from senior managers to secretaries, would undergo a detailed review process where their performance was judged by both managers and peers. They would then be ranked on a scale from 1 to 5. Those with a 1 received enormous bonuses. Those with a 5 got fired. Everyone in between was told they had to improve. The system was known internally as 'rank and yank'. It was exactly the kind of process that McKinsey had been advocating for in its talent reports.

In the aftermath of Enron's collapse, the journalist Malcolm Gladwell linked Enron, McKinsey and what he called the 'talent myth' which had gripped modern society. Gladwell set out some of the obvious problems with the obsession with talent: that in practice it's impossible to create systems that accurately judge someone's performance or potential; that focusing on talent prioritises the interests of a company's executives over those of their customers; and that cultures that excessively celebrate people because of their supposedly innate talents or intelligence cause psychological harm, whether in the arrogance and narcissism of those who are praised, or the self-loathing of those who aren't. In reality, he argued, an organisation's intelligence is a function of its systems, rather than the brainpower of individual employees.

Gladwell ended his article with a simple question: 'are smart people overrated?'[26] It's hard to avoid the conclusion that they had indeed come to be. By the 2000s, and despite all the controversies over intelligence, genetics and race which had been raging over the previous decades, the status and rewards granted to those judged as 'smart' seemed only to be growing. In the era of the knowledge economy, the idea of intelligence continued to reign.

Elite education

If the knowledge economy wanted 'talent', it needed an education system that would identify and nurture that talent. The old systems of higher education wouldn't cut it. Despite a big increase in access to higher education across the world from the 1960s, there weren't enough graduates to meet the demand for skills in the new knowledge economy. So universities at the turn of the century needed to grow.

But that created a new problem. If more people went to university, how could elite employers distinguish the best and brightest from the increasing mass of everyday graduates? What was needed was a separate tier of elite universities. In the age of globalisation, these elite universities in turn became global. If you attended one of these elite global universities you would be officially stamped as smart and hard-working, worthy of a career at the cutting edge of the knowledge economy. You would be anointed as part of the cognitive elite. If you weren't, then, well, best of luck.

A renewed faith in education, particularly elite higher education, became embedded in political culture from the 1990s. With the emergence of neoliberalism and the end of the Cold War, governments withdrew from previous commitments to securing full employment, welfare and basic services for all. In their place, they offered education-based 'opportunity'. In the era of globalisation all workers were competing in a global skills race, forced to participate in the global marketplace. To thrive in this global marketplace, everyone needed to develop the right skills. This meant embracing the opportunities offered by education.[27]

Governments increasingly saw the provision of these educational opportunities as their key task. Tony Blair famously said that his top three priorities for government were 'education, education, education', arguing that the British people needed the best education if they were to meet the challenges of globalisation. As well as trying to improve the school system, this implied getting more young people into universities. 'We think everybody ought to be able to go to college,' Bill Clinton argued, 'because what you can earn depends on what you can learn.'[28] These arguments were echoed by tech evangelists like Bill Gates, who claimed that technology would make educational opportunity available to everyone around the world for the first time, equalising opportunity and levelling society.

This, of course, wasn't entirely true. The flip side of governments providing educational opportunities was that workers were expected to seize those opportunities. 'What you can earn depends on what you can learn' was all well and good for those who were ready and able to learn the skills demanded by the knowledge economy. But what about those who, for whatever reason, couldn't learn, or couldn't learn as well as others?

This new ideology of educational opportunity wasn't just about learning. It was also about competition and credentials. It was, in part, a zero-sum game where there had to be losers in order for there to be winners. The opportunities available in the global knowledge economy were not limitless, as some of the globalisation boosters suggested. Those who fell behind in the educational race often found that there were few opportunities available, and that the knowledge economy left them with jobs that were less secure, less rewarding and less well-paid than what had come before. Added to this was the psychological and reputational harm suffered by those deemed not to have successfully embraced the educational opportunities available. The concept of the educational bargain implied that those who failed to grasp such opportunities *deserved* their lot. If you couldn't learn, why should you expect to earn?

The renewed global drive for education was reflected in a massive expansion of universities. The number of students in some form of higher education around the globe more than doubled between 2000 and 2020.[29] Enrolment rates increased across the world, often driven by targets set by governments convinced that a better-educated workforce was needed for their country to compete in the global knowledge economy. In 1999, for example, Tony Blair set a target of 50 per cent of young people going to university. It was a target rooted in contemporary language of intelligence, globalisation and the knowledge economy. 'In today's world,' he told the Labour Party Conference as he announced the target, 'there is no such thing as too clever. The more you know, the further you'll go.'[30] Today, just shy of 40 per cent of young people in Britain enter university at 18, up from less than 25 per cent in 2006.[31]

This expansion of higher education wasn't limited to Europe and North America. In 1990, less than 4 per cent of Chinese young people were in higher education. India sent the same proportion of young people to university as the United States had in 1920. But both systems were transformed over the course of the 1990s and the 2000s. By 2009, 24 per cent of Chinese young people were attending universities. The Indian student body rose from 4.5 million in 1985 to 17 million in 2010.[32]

This global expansion, though, progressed along two distinct tracks. One was the expansion of mass higher education. This saw an increasing proportion of young people entering university. They did so in large or rapidly expanding institutions, often offering shorter degree programmes or focusing on technical or vocational subjects. The other track took place within a much smaller set of elite global universities. Many of these institutions consciously decided not to expand in size. It's hard, after all, to claim that you're elite if you let too many people in. The combination of increasing demand for university education and a restricted number of places in the top universities meant that getting into these institutions became much harder. Up until the 1990s, admissions to elite universities weren't particularly competitive. But this changed dramatically over the following decades. In the early 1980s, Harvard and Stanford accepted around one in five applicants. Now they accept fewer than one in twenty. Between 1993 and 2019, the acceptance rate at the University of Chicago fell from 77 per cent to 6 per cent.[33]

Countries around the world aspired to copy the American model of the elite, highly selective research university.

Governments believed that this kind of university would produce the talent they needed to develop new technologies and to drive economic growth. The Chinese state separated off an elite group of universities from the mass provincial universities which were driving the expansion of student numbers. Universities like Peking, Tsinghua and Fudan were brought together into the C9 League (sometimes referred to as the 'Chinese Ivy League') and plied with government money with the aim of cementing their world-class status. This policy of creating world-class universities, particularly technical ones, was enthusiastically copied elsewhere, from the expansion of the Indian Institutes of Technology to the creation of the King Abdullah University of Science and Technology in Saudi Arabia.

As is the case with their Western peers, entrance into these global elite universities has become fiercely competitive.[34] China had revived its ancient tradition of national entrance exams by reintroducing the *gaokao* system after the Cultural Revolution. Taken by 13 million young people every year, the *gaokao* is a two-day series of tests that determines access to Chinese universities. Getting a place at any university is difficult enough, but competition for places at the elite C9 institutes is fierce. Students prepare for years with extracurricular study and private tutors, often sitting the test multiple times. The joint entrance exams for the Indian Institutes of Technology has been described as the hardest entry test in the world, with only one in forty-five applicants gaining a place. Despite the poor odds, applicants will often spend years preparing for the test, often leaving home and living in hostels while they attend special coaching centres.

The battle to get into elite universities has become so fierce partly because they are increasingly the only place that elite employers recruit from. As the idea of the 'war for talent' took hold in boardrooms across the world, recruiters increasingly shifted their attention to the top universities. It was here, they believed, that the best talent was to be found. 'The way that we win over our competitors is having smarter people in the firm,' one financial services HR executive told researchers. 'We are very keen to get the best talent, so we go after the elite . . . in other words, we want to get the top half percentile of university students.'[35] Another reported their major financial firm recruited from only eight global universities. Western companies were increasingly open to the idea that 'talent' could be found anywhere in the world, so were happy to diversify their recruitment beyond universities in their own countries. But at the same time they were limiting the range of institutions they hired from to only the most prestigious universities in each country.

The modern practice of restricting hiring to graduates of elite institutions had begun with companies like McKinsey in the 1950s, spreading across consulting, finance and other parts of the private sector from the 1980s. But it was soon increasingly evident in other sectors, including government. Over half of the appointments made by the early Obama administration had some connection to

Harvard. In his second term, two-thirds of cabinet members had attended Ivy League institutions, and over half had gone to Harvard or Yale.[36]

This trend, it has to be emphasised, didn't have anything to do with what graduates of these elite universities had actually learned. Partly it was a matter of convenience, reducing the number of applicants that firms had to sift through. But it was more about the status that attached to students who had been accepted into the most prestigious institutions. 'There are any number of places that do a pretty good job of teaching organization, accounting, finance, production and marketing,' the leading organisational scholar James March told Stanford Business School students in the late 1980s. 'They don't, however, do nearly as good a job of establishing that you're one of the very smart folks.'[37] What employers valued, in other words, was the selection process that students at these universities had already been through. The fact that places in these institutions were increasingly competitive and acceptance rates ever smaller meant, in the eyes of major employers, that graduates of these institutions must, by definition, be the best and brightest.

As education at elite universities became more important, the question arose of how students and employers could identify which institutions were elite, and which weren't. Luckily, help was at hand from a new technology with a key role in the modern rise of the cognitive elite: university rankings.

The university where I teach is currently very pleased with the fact that it has placed highly in one of the main global rankings. Students all around the world look at these rankings when they're deciding where to study. Many of my students are happy to get a place at our university because they feel it communicates something about their intelligence – that they have the mental ability to merit a place in one of the world's highly ranked institutions. Others, however, worry about it. They might have been hoping to get a place at one of very elite colleges that cluster in the top few places of such rankings. For them, coming to us is a second-best option, and they fear top employers will see the fact that they studied here as evidence of their second-rate abilities. Everyone who works in universities knows that these rankings are so subjective and arbitrary as to be more or less meaningless. But that doesn't stop the rest of the world paying attention to them.

All of this is a very modern phenomenon. University rankings had first been produced in the US at the start of the 20th century. Their origin was tied up with the history of intelligence, eugenics and psychology, all of which we will discuss over the following chapters. But it was only in the modern era of the knowledge economy and the cognitive elite that they came to possess the power and influence we see today.

The first ever university ranking was published in 1910 by the American psychologist James Cattell.[38] Cattell was a pioneer of mental testing and studies of intelligence, designing some of the very first mental tests used in the United States. He was also a eugenicist, who was deeply worried about what he believed was the declining genetic quality of the American population. Like other early intelligence scientists, he was interested in the study of 'eminent' men, and published a study of such men in his 1906 book, *American Men of Science*. This book provided the basis for Cattell's university ranking. The ranking system calculated how many eminent men of different ranks worked at each university. It combined this with data about an institution's reputation and the number of doctorates it produced to create a ranking of universities' 'scientific strength'. (Harvard was top of the list, in case you're wondering.) Cattell hoped that his list would help inform the choices of students, promote excellence among US universities in the scientific battle against countries like Germany and boost the influence of researchers within university administration.

Cattell's ranking didn't receive a huge amount of attention at the time, nor did the various alternative versions published over the following decades. The US government tried its hand at ranking research universities during the Cold War, when the massive expansion of scientific funding in the aftermath of Sputnik prompted concerns that public money needed to be channelled into the most effective scientific institutions.[39] But it wasn't until the 1980s that the university rankings 'boom' really began in earnest.

It was sparked by the magazine *U.S. News & World Report*, which published its first rankings in 1983, capitalising on the expansion of higher education and the demand for guidance from prospective students and their parents. This was the era that saw the emergence of the consumer movement. Consumers of higher education, particularly in the United States, where college fees were spiralling, were beginning to demand the same kind of information and guidance as consumers of cars and washing machines. The first surveys were entirely based on reputation, as measured in surveys of college presidents. Those same presidents immediately asked the magazine to stop publishing the rankings, alarmed at the potential impact on their recruitment and what many saw as the 'unfair' results they published. But the rankings, it turned out, sold a lot of magazines, so they were here to say.

The *U.S. News* rankings spurned imitators around the world. The first big global university ranking was published in 2003 by Shanghai Jiao Tong University. This favoured Cattell-style markers of status, such as the number of Nobel Prize winners linked to each university. The Shanghai rankings were quickly followed by other global university rankings from *The Times Higher Educational Supplement* and others. Today, each ranking system works slightly differently. But most are built around a similar combination of measures as those used since the early 20th century: things like reputation, research output, selectivity (judging universities as better if they are harder to get into) and graduate earnings. Rankings of business schools have proved particularly popular, offering a benchmark for the firms like McKinsey and Enron that wanted to ensure that they were recruiting only the very best MBAs.

These rankings have had a huge impact on the global education system. They're the obsession of many university leaders, whose pay and bonuses are increasingly linked to their ability to boost institutions up the league tables. They've also been embraced by governments, who often use them to help decide which universities should get most funding. Many countries outside of western Europe and North America have explicitly set goals around creating 'world-class' universities, as measured by the rankings.

This is certainly not what university rankings were originally designed to do, and the imprecision of their models often causes real problems. In 2005, for example, the Malaysian government celebrated when the University of Malaysia was ranked 89th in the world by the *Times Higher Education World University Rankings*. It turned out, though, that there were problems with the data, and when they were corrected the next year the university dropped to 169th. The Malaysian media announced a national education crisis, and the vice-chancellor was sacked. Elsewhere, universities around the world have been caught trying to cheat the rankings, whether by submitting dodgy data or deliberately bad-mouthing their nearest competitors in reputational surveys.[40]

Ultimately, rankings proved popular because they offered a way to identify elite universities, and getting into these elite universities increasingly represented the best, often the only, path to professional success. The question remained, however, what exactly distinguished the graduates of elite universities from everyone else? What made them better, more employable, more professionally desirable?

The most common answer to that question in recent decades has been not talent or intelligence but *effort*. As Daniel Markovits has argued in his study of modern meritocracy, the rise in economic inequality over recent decades has been combined with a cult of extreme work among elites.[41] This cult can be seen in everything from tech CEOs boasting on social media about how they start their working day at 5am to young financial workers being driven to suicide by extreme stress and overwork. This overwork is partly functional: elite workers believe extreme hours are necessary to do their jobs properly, and managers claim they are indispensable for their business's success. But it is also performative: hard work done with the intention that others can identify you as a hard worker.

This performative cult of effort became a way to justify the fundamentally unjustifiable economic rewards which have accrued to elites over recent decades. People earning the highest salaries in tech, finance or consulting felt the need to justify those salaries, both to themselves and to others, on the grounds that they *deserved* their rewards. To argue that such rewards were simply the product of natural intelligence would imply that they were partly a matter of luck, a consequence of the lottery of birth. They were much easier to justify, on the other hand, if they were framed as the result of someone's grit, determination and hard work.

But while the importance of intelligence to the meritocratic race was sometimes denied, more often it appeared *in combination* with things like hard work and ambition. To succeed in education and in life, according to this view of the world, required both brainpower *and* exceptional effort. As Barack Obama put it in a speech about higher education in 2014, 'ultimately what matters . . . is making sure that bright, motivated young people . . . have the chance to go as far as *their talents and their work ethic and their dreams* can take them' (my italics).[42] Sometimes this meritocratic cocktail of intelligence, motivation, effort and ambition was even more explicitly linked together. According to one young American investment banker, the hyper-successful Ivy League graduates in his industry perceived anyone who chose a lower-paid career with a better work-life balance as 'less smart'. In other words, choosing to work hard was a sign of high intelligence.[43]

In the post-*Bell Curve* era, the celebration of educational achievement was normally couched in the softer language of 'brightness' and 'smartness' rather than the tarnished language of IQ. And at a time where elites felt under pressure to justify their ever-increasing wealth, success was usually credited to a combination of intelligence and hard work rather than to brainpower alone. But higher education, particularly elite higher education, had clearly come to enjoy a new status and purpose in the age of the cognitive elite. It was the thing that marked young people out as 'one of the smart ones', those destined for – and deserving of – professional and economic success.

Neoliberal supermen

Not all members of the modern cognitive elite are equally elite. The largest group are the lower-paid knowledge workers – those with university education and professional, white-collar careers that give them a degree of social status and economic security without a huge amount of wealth. Above them are a narrower category of those who attended elite colleges and universities and who went on to staff the higher echelons of the best-paid industries – senior figures in technology, finance, consultancy and so on. These people enjoy both status and material rewards rooted in their educational achievements and the nature of the work they do.

But above this stratum are a much narrower group which have come to symbolise today's cognitive elites. We can label them in different ways: the billionaire (or almost-billionaire) class; the super rich; the 0.001 per cent; the company founders, entrepreneurs and inventors. Whatever we call them, they are the people who stand at the top of the hierarchy of power, money and status in modern society. I am going to call them the neoliberal supermen.

The status of these men (and they are men almost to a man) offers an eerie echo of the late 19th century, when psychologists and eugenicists like Francis Galton lauded the 'great men' of the era and tried to uncover the physical and psychological secrets of their greatness. The celebration of gifted great men appeared to fall into abeyance in the middle of the 20th century. But it has returned with a vengeance in recent years.

The idea of intelligence is central to the image of the modern neoliberal superman. They are people who generally benefitted from elite education. They and their acolytes ascribe their success to a range of factors, including hard work, which, as we have seen, has become so important to elite identity. But a core part of the myth surrounding these figures is their hyper-intelligence, the idea that their success is due to their unique set of mental skills, their creativity or their extraordinary brainpower.

There are plenty of examples we could point to, but the most high-profile neoliberal superman of the modern era is undoubtedly Elon Musk. Musk's boosters, like the podcaster Joe Rogan, literally describe him as a 'superhero', someone gifted with a 'radical sort of creativity engine' that is advancing human civilisation.[44] According to the *Iron Man* screenwriter, it was this image of Musk that inspired the Tony Stark character in the Marvel cinematic universe. Musk sees himself as someone whose mind works differently to everyone else's, both in the sense of being neurodiverse and of operating on a higher mental plane, referring to the way he uses his brain to 'compute'. He values similar traits in others, as seen in his 2024 call for 'super high-IQ small-government revolutionaries' to staff his new Department of Government Efficiency. As his biographer Walter Isaacson noted, the flip side of his celebration of genius is the frequency with which he uses the word 'stupid', attacking those around him as 'morons' and 'idiots' when he gets into fights, just as his father would do to him when Musk was a child.[45]

The neoliberal supermen like Musk stand at the apex of modern society. We have created a social and economic hierarchy headed by a small number of billionaires who are celebrated, by themselves and by others, as hyper-intelligent innovators. But there was nothing natural or inevitable about this process. It is a myth, a cult, which has emerged from the collision of a number of different historical forces.

The first of those historical forces is the emergence of the modern image of the entrepreneur. The riches of the neoliberal supermen are attributed to their status as entrepreneurs, often serial entrepreneurs, self-made men whose business savvy allowed them to bootstrap their way up from humble beginnings.* Today, entrepreneurship is seen as one of the cardinal social virtues, taught in schools, lauded by governments and celebrated on social media. The idea of the entrepreneur is over a century old, but this popular celebration of entrepreneurship is a more recent phenomenon.

In the early 20th century, the new scientific management theories of men like Frederick Taylor saw management as a matter of technical expertise. The men tasked with running the era's cutting-edge industrial firms were encouraged to think in terms of production processes,

---

* The reality in almost every case is that those beginnings weren't particularly humble – almost all of the neoliberal supermen come from privileged backgrounds of various sorts, often tapping into the money of family members or networks to fund their early careers. But why let truth get in the way of a good origin story?

time and motion studies, and business efficiencies. But other management thinkers in the US and Germany reacted against this ideal. They worried that uninspiring technical managers would breed uninspired, alienated workers who didn't identify with the interests of the businesses they worked for, creating a culture of resistance to innovation. Instead, they advocated for entrepreneurial management, emphasising leadership and zeal over rationality and technical skill.[46]

The ideal of the entrepreneurial manager lost some of its lustre in the middle of the 20th century. This was a period when corporations and bureaucracies grew massively in size, and enthusiasm for 'planning' swept the world of business. Being a manager within a modern, massive conglomerate came to be understood as a matter of performing specific tasks to ensure that the organisational machine continued to function smoothly, rather than showing initiative or taking risks. Economists like Joseph Schumpeter argued in the 1940s that capitalist progress in the past had been driven by an elite class of entrepreneurs, but that this class was being forced out of economic life by the giant corporations of the modern era. The innovation and creativity of small and medium-sized firms was being replaced by the administrative efficiency of commercial bureaucracies.[47]

But the 1970s witnessed the beginning of a revival of entrepreneurship.[48] It was a revival driven by a group of economists in Germany and elsewhere, many of them linked to the neoliberal Mont Pelerin Society. These economists argued that entrepreneurs were key to the healthy functioning of markets, discovering and trialling strategies to take advantage of new opportunities. They were particularly interested in promoting the public image of the entrepreneur. This task was taken up with enthusiasm by the new generation of free market think tanks that were emerging in Britain and the United States at the time. Universities and business schools were also eager adopters, ramping up the teaching of entrepreneurship from the 1970s and 1980s. All of this proved hugely successful in raising the profile and status of entrepreneurs in modern society.

There was a crucial difference, though, in the way neoliberal economists of the 1970s and 1980s understood entrepreneurship compared to both earlier and later uses of the term. Entrepreneurs, the neoliberal reformers of the 1970s believed, were not a narrow elite. Rather, entrepreneurial talent was everywhere, available in almost unlimited supply. Intelligence had little to do with this understanding of entrepreneurship. If being an entrepreneur was about sensing and pursuing market opportunities, it didn't require any particular technical skills or mental ability, just a sense of how the world was working and the boldness of character needed to exploit it. Entrepreneurs were already working in all areas of the economy. But their talents were either being restricted by petty rules or wasted in big bureaucracies. The task of governments was to set this huge pool of entrepreneurial talent free.

In order for this neoliberal vision of the entrepreneur to coagulate into today's cult of the superintelligent billionaire, it had to be combined with another current of right-wing ideas about biology and IQ. As discussed later in the book, the controversy surrounding *The Bell Curve* in the mid-1990s harmed the image of IQ and intelligence science, and prompted a backlash against those who argued that inequality was rooted in genetics. But the rejection of these ideas by mainstream liberal society didn't mean they had disappeared entirely. On the right, such ideas survived, and even flourished.

As the historian of neoliberalism, Quinn Slobodian, has recently shown, a new strand of thinking on the right and far right emerged from the 1980s.[49] Whereas in the past those on the right had rooted their arguments about inequality in appeals to religion or rationality, they now began to appeal to biology, psychology and genetics. In particular, they appealed to a genetically based understanding of intelligence to argue that economic inequality was rooted in nature, and that any attempts to ameliorate it were doomed to fail. This shift was driven by a coalition of psychologists, free market think tanks, and right-wing intellectuals. They were, paradoxically, boosted by the *Bell Curve* controversy, which drew public support from influential figures like the far-right US presidential candidate Pat Buchanan. All of this helped to drive a revival of IQ on parts of the right.

It was a revival that tapped into some of the major intellectual trends of the 1990s and 2000s. One of these was the notion of the knowledge economy, with its celebration of symbolic analysts and brain workers. Another was the rise of neuroscience, driven by new scientific studies of the brain. These studies inspired popular and political enthusiasm for harnessing knowledge of the brain to improve society. George H.W. Bush, for example, declared the 1990s to be the 'Decade of the Brain', aiming to boost the profile of brain research and its potential to contribute to everything from mental health treatment to the war on drugs. The third was a renewed enthusiasm for genetics, inspired by the success of the Human Genome Project in 2003, and the various technologies to map and manipulate human genes that followed in its wake. This led to a scientific and popular turn back towards emphasising the role of genes in human differences, reversing the environmentalist trend that had dominated many fields from the 1960s.

The combination of these trends encouraged the belief that intelligence was crucial to the economic success of individuals and nations; that it was fixed in the mechanics of the brain; and that it was determined by nature more than nurture. These ideas offered a convenient platform for various right-wing talking points. Charles Murray, the political scientist who coauthored *The Bell Curve*, argued that differences of sex, race and class are rooted in biology, and particularly in differences of brain function and cognitive ability. In other words, that economic and social inequality is rooted in biological inequality. Any political or social policies trying to promote equality in these areas was, he believed, inevitably constrained by these biological facts.[50] Jordan Peterson has argued that 'there's no place in our cognitively complex society for 1 in 10 people' because of their low IQ.[51] Other right-wing theorists in the US and Europe have gone even further, arguing that global inequality is caused by the varying IQ levels of nations and races, and that immigrants should be given IQ screenings.

This IQ revivalism, once marginal, has entered the mainstream with the global rise of the far right. As a tradition it explains why people like Donald Tusk and Elon Musk are so quick to reach for the language of IQ, in a way that feels like a throwback to the pre-*Bell Curve* past. As we will see so often in this book, the appeal of the idea of intelligence for the new right lies in the way it can be used to justify inequality and hierarchy. Their arguments draw on a long history of ideas about intelligence which remain embedded in our culture today, and which make their claims sound like common sense to many.

The modern figure of the neoliberal superman, then, is rooted in both the increasing celebration of the entrepreneur and the right-wing revival of intelligence science. It is a modern cult, a vision of the world embraced by fans and flunkeys of the billionaire class, and by the neoliberal supermen themselves. And it's a vision that manifests itself in some strange ways.

One of those is the obsession many of the neoliberal supermen have with population and birthrate. Elon Musk, to return to our previous example, has spoken frequently about his concerns that declining birthrates threaten the future of humanity. He has certainly practised what he preaches. At the time of writing he appears to have fathered at least fourteen children with four women (plus the help of various surrogates), and has made some of his children a very visible part of his public life. But, as one of the mothers of his children reported, Musk doesn't just want everyone to have more kids; 'he really wants smart people to have kids'.[52] Musk isn't alone in holding these ideas. Pavel Durov, the Russian founder of the messaging app Telegram, has fathered over 100 children in 12 different countries through sperm donation, motivated by his desire to meet the need for 'high-quality donor material'.[53] He's also offered free IVF treatment for women at a Moscow clinic who agree to use his sperm.

As we will find in the coming chapters, these kinds of ideas have a long lineage. Concerns about the size and 'quality' of the human population were central to the development of the eugenics movement in the early 20th century. Many early intelligence scientists shared these concerns, and the modern idea of intelligence played an important role in eugenics debates about population quality. These ideas were often tangled up with arguments about racial differences in IQ. Some of these ideas now seem to be trickling down into society at large. A new generation of high-end fertility clinics is now offering 'elite' sperm from men with desirable character traits and abilities. Many of these clinics specifically cater to those who want to produce intelligent children, marketing donors as 'the smartest guy in the room', or highlighting their 'Einstein-like math skills'.[54]

The cult of the neoliberal superman is also reflected in the world of philanthropy. The current billionaire class, especially those from the world of tech, have been particularly interested in so-called 'effective altruism'. This ideology, which emerged from a group of philosophers and academics in Britain and the US in the early 2000s, encompassed a range of ideas, from the fairly sensible ambition to give money to causes that actually make a difference, to a set of odder and much more abstract concerns about existential risks to the world, the long-term survival of humanity and the moral value of future generations. Its appeal to tech elites lay partly in its hyper-rational, utilitarian, quantified approach to philanthropy. But there were other obvious reasons why these tech elites might be attracted to the movement's 'earn-to-give' philosophy. This held that the best way to have a positive impact on the world was to make a huge amount of money, and then to give it away.

In the early 2020s, Sam Bankman-Fried became the poster boy for effective altruism. A billionaire who shot to fame as the head of the crypto exchange FTX, Bankman-Fried had previously worked at an NGO called the Centre for Effective Altruism. FTX, he claimed, was created as a vehicle to earn money that could then be given to causes that would change the world for the better. It was a promise that helped the scruffy young entrepreneur working in a notoriously shady industry to attract fawning attention from titans of industry and global political leaders. To help achieve his goal, Bankman-Fried set up the FTX Foundation Future Fund in early 2022. It was led by some of the founders and leading lights of the effective altruism movement, and promised to give away between $100 million and $1 billion in its first year.

The fund's work suggested a particular obsession with intelligence, elites and the transformative power of what it called 'exceptional' people. Its stated goals included 'empowering exceptional people', supporting and identifying the most talented young people around the world to fulfil their potential and go on to help solve humanity's most pressing problems. Sometimes this faith in the power of the intelligent and exceptional was rather vague. In its work on the risks of AI, for example, the fund aimed to increase the number of 'smart' people working on machine learning and AI policy because 'we think it's a good bet that these people will find ways to help in the future'.[55] Others were more concrete. In the first five months of its existence, it issued 18 grants totalling $10 million under its 'Empower Exceptional People' funding stream. One of these grants was given to the Atlas Fellowship, which ran summer schools for 'exceptional students' around the world, teaching them about global challenges and encouraging them to think about existential risks to humanity. Another project, run by a recent Cambridge graduate, promised to identify 'top young students around the world' and support them to work on 'improving humanity's long-run prospects'.[56]

The effective altruism crowd always had a tendency to celebrate an elite of the rich and the rational, seeing them as the route to saving humanity. Bankman-Fried, like other neoliberal supermen, was widely lauded for his intelligence and seemed to have a high regard for his own abilities. It is perhaps unsurprising that a billionaire who sees himself as an intelligent person improving the world will conclude that the best way to improve the world further is to find and support more intelligent people who can continue the cycle. This was a philanthropy that put the idea of the neoliberal superman and faith in cognitive elites at its heart. As will be demonstrated throughout this book, beliefs like this aren't new. From Francis Galton's celebration of 'great men' in the 19th century to the creation of Mensa and the emergence of the gifted and talented movement, people who think they are intelligent have long believed that intelligent people are the key to progress.

In the case of the FTX foundation, however, things didn't quite go to plan. Bankman-Fried, it turned out, was a crook. Following the dramatic collapse of FTX, he's currently serving a 25-year prison sentence for what amounted to one of the biggest frauds in financial history. The FTX foundation and its Future Fund collapsed at the end of 2022, having only paid out a fraction of the billions of dollars Bankman-Fried had promised. Much of that money has since been clawed back to compensate FTX's investors. The effective altruism movement, already the subject of widespread criticism, seems unlikely to recover from the blow.

At the time I'm writing this, in early 2025, Donald Trump is a few months into his second term in office. One of the features of these early months has been his administration's attack on DEI – diversity, equality and inclusion. The new American right, eagerly aped by their fanboys and fangirls abroad, have denounced DEI programmes as a fiendish Marxist plot to undermine Western civilisation (much to the surprise of Marxists everywhere). Trump has issued an executive order terminating all DEI programmes in government agencies and suppliers, describing them as 'illegal and immoral'.[57]

In its place, figures like Elon Musk have endorsed a new concept: MEI, or merit, excellence and intelligence. The term was coined in 2024 by an AI company founder (now Meta's chief AI officer) named Alexandr Wang.[58] Wang argued that his company would only succeed if it was a genuine, talent-based meritocracy. 'That means we hire only the best person for the job,' he wrote, adding that they 'seek out and demand excellence, and we unapologetically prefer people who are very smart.' As American corporations have been enthusiastically rowing back from their previous public commitments to diversity, MEI has been widely touted as a radical new solution to society's ills.

It is, of course, no such thing. In reality, it's the culmination of over a century's worth of obsession with the notion of intelligence, and of the more recent faith in the idea of a cognitive elite which has taken root since the 1990s. The most powerful political figures, the leading industries and the richest men in the world are publicly embracing the idea that society should be organised to ensure that only the most intelligent rise to the top.

Our current era, then, is shaped by the idea of intelligence. Our job now is to find out where that idea came from.

2

Intelligence Before IQ

> 'That the word *intelligence* describes something real and that it varies from person to person is as universal and ancient as any understanding about the state of being human ... Gossip about who in the tribe is cleverest has probably been a topic of conversation around the fire since fire, and conversation, were invented.'
>
> Richard Herrnstein and Charles Murray,
> *The Bell Curve*

THIS BOOK EXPLORES THE history of intelligence since the end of the 19th century. And there's a particular reason it begins at that point: because before the 19th century, intelligence didn't exist.

Now, I accept this might seem a bit counterintuitive. Whatever we might think about what intelligence means, we tend to assume that it's a real existing *thing* – that it's a fundamental characteristic of human beings, a bit like eye colour, or temperament. It would make no more sense to say intelligence didn't exist in the past than to claim that height didn't exist in the past.

We also tend to think of it as *ahistorical*. By that I mean that we assume a person living five hundred, a thousand or perhaps even ten thousand years ago would have been able to identify the people around them who were intelligent in the same way they would have been able to identify those who were tall, or timid, or good at hunting. This is the kind of idea that the quote at the start of the chapter is getting at.

But is this really the case? It turns out the answer's not as straightforward as we might think.

Part of the difficultly in knowing how intelligence was understood in the past is the language that was used to describe it. 'Intelligence', in its modern meaning of mental ability, is a fairly modern term. Up until the 18th century the word was mainly used to refer to information and its exchange, a meaning that lingers on in terms like 'military intelligence'. Other words (which we'll discuss in more detail later) were used in different languages to refer to mental abilities. From the 16th century, some of these terms encompassed the idea of high, general intellectual capacity. But this was only ever one of a range of meanings attached to them, and by no means the most common one. When people talked about intelligence in the past, they weren't necessarily talking about the same thing as we are when we talk about it now.

One of the biggest differences is that intelligence used to be understood in a much broader way than it is now. Today, the way we think about intelligence is quite narrow, focused on mental abilities linked to verbal and spatial reasoning, memory and the ability to learn. For much of human history, however, discussion of these mental abilities was tangled up with lots of other ideas about character and temperament.

In many cultures at many different times, for example, you couldn't be considered intelligent without also being wise, temperate, kind or trustworthy. In many places you were also unlikely to be considered intelligent if you weren't (a) rich and (b) a man (although this is maybe an area where our modern ideas haven't changed that much). Perhaps the most striking difference, however, is that intelligence was often understood as something that was in some way connected to the divine, something that encompassed both the human mind and the cosmos. Human intelligence was understood as something that mirrored or groped towards divine intelligence. It was something that had the potential to bring people closer to God.

So when I say that before the 19th century intelligence didn't exist, I mean that our modern narrow, secular understanding of intelligence didn't exist. If we could travel back in time five hundred, a thousand years or ten thousand years, the people we met wouldn't necessarily be able to identify those around them who were 'intelligent' in the same way they would be able to identify those who were tall, or kind, or good at hunting, because the way they thought about intelligence would probably be fundamentally different from our own.

Many of the intelligence scientists of the 20th century would claim that, whatever people believed about intelligence in the past, intelligence *as a thing* still existed. Indeed, one of the favourite pastimes of early intelligence researchers was to try to calculate the IQs of famous historical figures, using historical sources to work out, for example, if Napoleon was more intelligent than Mozart, or Newton cleverer than Shakespeare. But it's not obvious that this idea of intelligence as something 'universal and ancient', as the authors of *The Bell Curve* put it, is really true. Intelligence isn't a single objective characteristic like height; it's a label that has been attached to a range of mental abilities. Before the 19th century, neither the label itself nor our understanding of that specific collection of mental abilities existed in the same way.

Something changed, then, around the end of the 19th century. It was in this period that our modern ideas about intelligence crystallised and began to take on more importance. In other words, this was the point in time when intelligence was *invented*. But we didn't just conjure intelligence out of thin air. Rather, we built it upon some very ancient foundations. For thousands of years, people have been thinking about the nature of the human mind, about its abilities, and about why those abilities matter. This history is where our modern ideas about intelligence emerged from.

In this chapter, then, we need to think about the prehistory of our new era, about how intelligence and its significance were understood before the late 19th century. As we'll see, these were ideas that were always bound up with the question of human inequality and hierarchies.

Reason and the philosophers
In the beginning, the mysteries of the human mind were a subject for the philosophers. It was the philosophers of the ancient Greek and early Islamic worlds who laid the foundations on which our modern ideas about intelligence were built. That's not to say that the Greek and Arab philosophers trying to uncover the secrets of the human mind were talking about 'intelligence', at least not in our modern sense of a general measure of mental ability. But their debates about understanding, learning and calculation provided some of the ingredients that would later be incorporated into our notions of what it means to be smart. And one concept in particular was fundamental to the way intelligence would later come to be understood: that of *reason*, the faculty through which we grasp truths, attain knowledge and work out how to live well.

The nature and role of human reason was one of the most important problems for the philosophers of ancient Greece. The concept of *nous* – commonly translated as 'mind' or 'intellect' – was central to the metaphysics of both Plato and Aristotle. *Nous*, the Greeks believed, was the highest form of human mental activity, the rational component of the human soul which allows us insight into the true essence of things. But, like most historical visions of human intelligence, it also had a transcendent, even spiritual quality. It was a godlike ability through which humans were able to grasp fundamental truths and forms.

The Greeks undoubtedly thought that human reason was important, but they refused to treat it in isolation. Plato, for example, used the metaphor of a charioteer to understand the nature of the human soul. The charioteer represented something approaching human reason or rationality, giving drive and direction to the human soul. But to make this happen, the charioteer has to constantly struggle to control two different horses, one representing moderation and honour, and the other our unruly passions and desires. Reason was central to human existence, but it was inseparable from our passions, our values and our character.

The Greeks were also careful not to treat reason with more importance than it deserved. We tend to associate ancient Greece, particularly Athens, with calm rationality and deliberation, imagining groups of bearded philosophers sitting around under olive trees in white flowing robes, patiently unravelling the mysteries of the human soul and human society. But this emphasis on rationality shouldn't be overstated. In Greek myths and literature, the cold powers of reason were far less celebrated than the 'cunning' or 'guile' (*metis*) that was displayed by many of the most famous characters – be they gods or humans.[1]

Philosophers such as Plato were also interested in the role that reason – and those endowed with greater abilities to reason well – should play in human societies. In *The Republic*, Plato argued that the ideal state would be run by philosopher-kings, who would save human society from barbarism by combining wisdom with political greatness.[2] Plato's idea of what constituted a philosopher certainly wasn't reducible to the modern idea of intelligence – it encompassed things like virtue and the love of truth and justice, which have largely been separated from the modern sense of the word, but there were some similarities. Philosophers, he felt, were people who demonstrated a love of learning from an early age, who had a good memory, who could apply themselves to intellectual tasks and could acquire new ideas easily, and who perfected their abilities through years of study. He argued that some people were born to be philosopher-rulers, and others were born to be ruled. He also advocated selective breeding between the 'best' citizens in order to strengthen the natural qualities of the state and its people. As we'll discover later in the book, these were ideas that appealed to many of those who embraced the idea of intelligence in the 20th century.

Like Plato, Aristotle believed that humans were divided between those born to rule and those born to be ruled over. Natural rulers were characterised by their mental powers, natural slaves by their physical powers. As well as mental differences between individuals, Aristotle also thought there were fundamental differences between societies. The psychological characteristics of different peoples, he believed, were linked to the physical environments in which they lived. The cold countries of northern Europe produced people of low skills and understanding, but with a hardy sense of political independence. The peoples of hotter regions in Asia possessed more understanding, but lacked spirit and were therefore more naturally subjects or slaves than rulers. The Greeks, inhabiting a happy climatic median, combined both spirit and reason. And as luck would have it, this made them born to rule over others.[3] This idea that the human mind and the physical environment were somehow linked was shared by other ancient thinkers, like Galen and Ptolemy. It was an idea that would reappear in some of the major debates about intelligence and race that exploded into life in the 20th century.

So the ancient Greeks pioneered the study of reason. They understood it as an important part of the human existence and of human society, albeit one that was impossible to separate from our other characteristics and virtues. Perhaps more importantly, they linked it to ideas about human inequality. Some people or groups of people, they believed, possessed high levels of reason or other mental abilities and this justified their rule over everyone else who lacked such qualities.

If reason was of philosophical interest to the Greeks, the early Islamic Arab world embraced the values of reason, intellect and learning in even more profound ways. Ideas about human reason and intellect characterised the Islamic 'Golden Age', the intellectual revolution that took place during the early centuries of Islam and under the Abbasid Caliphate (750–1258). Across the Arab world, particularly in cities like Baghdad and Basra, the arrival of Islam launched a new age of philosophy and science, driven by faith that reason and imagination could uncover the secrets of human and divine truth. Thinkers such as Avicenna and, later, Ibn Khaldun made huge strides in philosophy, astronomy, chemistry and mathematics, laying the foundations for the European Renaissance and Enlightenment in later centuries. The extent of Islamic learning during this period is difficult to overstate. At a time when the largest libraries of France or England held only a few thousand titles, libraries in Muslim Spain had collected over a million.[4]

This Islamic age of reason has sometimes been attributed to the influence of Greek and Roman philosophical texts that were translated and preserved in the Muslim world. It's true that the Greek influence was important, as were texts from Persia and India. But the turn towards reason, science and philosophy was driven much more by the emergence of Islam itself. The new religion inspired believers to seek after truths about the nature of God and the Qur'an, of justice, and of free will. Often this search was based on a close study of the holy texts, and human reason was understood as key to unlocking their secrets. The textual basis of Islam, alongside the advance of new paper technologies, drove a huge expansion of literacy and writing cultures.

The nature of human reason and intellect was also one of the key questions for philosophers in the Arab world. Al-Farabi ($c.870$–$c.950$), for example, was one of most important early Islamic philosophers, known as the 'Second Master' after Aristotle. He was deeply interested in human psychology and reason. In his *Treatise on the*

Intellect and *The Virtuous City* he tried to map out the human intellect. He argued that human rationality could be both practical and intellectual, but that it was also tied up with our imagination. Intellect was something that all humans possessed, he believed, the highest faculty of the human soul and the key to happiness. But it was also something that could be developed and perfected. By developing our intellect, everyone was capable of attaining higher levels of understanding and truth.

Like Plato and Aristotle, the Arab philosophers like al-Farabi understood intelligence as a divine phenomenon as much as a human one. Intelligence was associated with God, the angels and the immaterial world. In the hierarchy of the cosmos, the lowest forms of this divine intelligence could correspond or unite with the human intellect.[5] That didn't mean that humans could possess divine intelligence. But it did suggest that those humans who mastered the art of abstraction or the deepest understanding of the human soul could approach a knowledge of the immaterial world. And ultimately, this knowledge was the reason that intellect was so valuable. Beyond the pursuit of science, medicine and technology, intellect and reason were the tools that enabled humans to understand God and to untangle fundamental theological questions about virtue, truth and justice.

Chinese meritocracy
The Greek and Arab philosophers led the way in exploring the nature of human reason and its role in human societies.

But these ideas had a limited impact on the societies and political systems of their day. Plato might have dreamed of philosopher kings, but philosophers remained philosophers and kings, by and large, remained merely kings.

It was in China that these ideas were really put into practice. The Chinese civil service examination system was the first systematic attempt to construct, at least in theory, a form of intellectual meritocracy where the qualities of a person's mind mattered more than the family they were born into. It had such a impact on everyday life that, within a few centuries, young men were copying the entire text of Chinese classics in tiny characters onto their underwear in a desperate attempt to get ahead in life by cheating on their exams. A brave new world had dawned.

The examinations had an incredibly long history. Their origins lay in the Eastern Han era (25–220 CE). They were developed under the Song dynasty (960–1279). And they reached their greatest extent under the Ming (1368–1644) and Qing (1644–1911) dynasties. They were only abolished in 1905, just a few years before the end of the Chinese imperial era.

The scale of the examination system in imperial China is also hard to overstate. At its peak in the late Qing era, around 5 million people were sitting for the examinations every two years.[6] They were held in over a thousand Chinese counties, with an elaborate system of tiered examinations rising from local to municipal levels, and culminating for the select few in metropolitan and palace examinations in Beijing. Successful degree holders had the right to apply for positions within the Chinese government system.

Each of these examinations represented a vast investment of time, money and effort. Candidates spent years memorising the required texts and mastering the necessary arts. Examinations themselves could take three weeks to complete, with candidates required to travel to special compounds where they would be sealed off from the outside world. They were given a few days to write each essay, and would then wait for it to be graded and ranked. The Nanjing compound alone could accommodate 17,000 candidates. Candidates would often travel with their families, and the areas around the compounds would be surrounded during examination season by merchants, food vendors and entertainers catering to the needs of the excited crowds.

Elaborate security systems were put in place to prevent cheating, with guards stationed on rooftops and in alleyways. But the scale and importance of the examinations meant cheating could never be entirely prevented. People tried to impersonate candidates, passed essays into the compound and bribed examiners. Others used the aforementioned trick of smuggling the classical texts into the compound by writing them out in tiny characters on their underwear. The stress of the examinations would often drive candidates into illness and depression, resorting to fortune tellers for guidance and reassurance. And overworked examiners often only had time for one or two words of feedback, raising complaints that marks were inaccurate or arbitrary.

The examinations provided a stern test of a candidate's mental abilities. They were based around 'Way Learning', the Neo-Confucianism rooted in classical Chinese texts. Candidates were expected to memorise the *Four Books* and *Five Classics*, often alongside other dynastic histories. This could involve memorising over 400,000 characters. In the examinations they had to use that knowledge to write elaborate, rigidly constructed 'eight-legged essays', while demonstrating their mastery of the art of calligraphy. Candidates were also asked to master legal codes, write poetry and produce policy essays.

None of this could be attempted without years of study, and even then only a small minority would succeed in being appointed as government officials. The system ensured that there was an influx of new ambitious, classically educated officials into government service every few years. Those candidates who failed to gain appointments formed a literate, highly educated cultural elite that fundamentally transformed Chinese society. The Chinese system also inspired similar models in neighbouring states, such as Korea and Vietnam.[7]

To our modern eyes, it's very easy to see these massive systems of standardised, timed, academic exams as something akin to today's education system, designed to identify those who are most intelligent, talented and driven. But while the idea of identifying those with the highest abilities to run the state was important to the Chinese system, we need to be wary of seeing the examinations as purely a test of brainpower.

Just as the Greek and Arab philosophers had understood reason and intellect as inseparable from virtue and character, the Chinese system was designed to test both mental and moral abilities. The examinations' focus on Way Learning had emerged under the Song dynasty as a way of reimposing conservative values. Mastery of these texts and ideas was understood as a test of moral, as much as intellectual, worth. Through studying the classical texts, candidates were understood to be undergoing a process of spiritual self-cultivation. Those who succeeded in the exams were therefore the ones who demonstrated the most commitment to self-cultivation.[8] It was this focus on values and self-cultivation which maintained the classical texts at the heart of the examination system for so long, excluding technical or scientific subjects like law or medicine. The exams were as much about identifying values and morals as skills and talents.[9]

We should also be wary of seeing the examination system as genuinely meritocratic. Those who enthusiastically champion the idea of meritocracy today sometimes offer uncritical histories of the Chinese system which overstate its impact. It certainly created far more opportunities for social mobility than in Western societies of the era. Chinese culture was full of stories about the lowly sons of peasants triumphing in the examinations and propelling their families into the ranks for the elites. But the system was definitely not designed to tap into the talents of *everyone* in Chinese society. Large parts of society were formally excluded from examinations, including all women, those in 'unclean' occupations and the clergy. Much of the rest of the population were *de facto* excluded by the requirements for years of schooling and mastery of classical Chinese. At a time when there was no public school system, this was impossible for the children of the semi-literate masses to afford. In effect, then, the examination system was the preserve of the top 10 per cent of society – the gentry, military and merchant families. The social mobility that did occur was between the lower and upper orders of these elites.

And even beyond the examinations, the Chinese commitment to meritocracy was not always particularly strong. Although success in the examinations provided opportunities to attain official roles, it was not the only route. By the late 19th century, for example, only a third of imperial officials were degree holders.[10] The rest were able to purchase their offices (an entirely legal process, with the price set by the Board of Revenue), or were appointed either through recommendation or as a member of one of the hereditary military orders. And all this, of course, took place within a system of hereditary imperial rule. The examination system allowed emperors to style themselves as 'sage-kings' or imperial teachers, sitting atop a vast system of scholarship. A notionally meritocratic, education-based appointment system ultimately helped to bolster a hereditary imperial system.

Despite these limitations, comparing the merit-based Chinese system to other states shows how remarkable it was by the standards of the time. While Chinese rulers were developing these elaborate systems to identify the most talented administrators across their vast territory, power and influence in feudal Europe was still based almost exclusively on birth. It was not until the 19th century that European states, partly inspired by the Chinese example, began to introduce competitive examinations for government posts. This was the first time that European feudal elites were forced to confront some of the same meritocratic pressures that had begun to undermine their Chinese equivalents in the 10th century.[11]

Things also looked very different in other Asian states of the time. In Tokugawa Japan, for example, attempts to adopt some of the meritocratic principles of the Chinese system were undermined by the more hierarchical nature of Japanese society and of the samurai class. When a new network of national schools was created in the 17th century, teaching a similar classical curriculum to those of China, elaborate rules had to be introduced to reflect the social status of different students. Sons of the highest families, for example, were allowed to attend school with one servant to look after their sandals and another to carry their umbrella. Those of middle-ranking families could have a sandal holder but not an umbrella carrier. And those of the lowest ranks, tragically, were denied the use of servants altogether. Even as a more competitive, meritocratic education system developed in the 18th and 19th centuries, it remained focused on providing individuals with the training appropriate to their hereditary status, rather than on promoting the most able and talented.[12]

By the end of the 19th century the Chinese examination system was increasingly under attack. Its critics alleged that it was unfit for the modern world, producing men adept at empty rhetoric but devoid of the scientific and practical expertise needed to manage a modern state. It was abolished in 1905, but its legacy lived on. In the West, the Chinese system offered inspiration for those who wanted to promote competitive, meritocratic education models. In China, elements of the system re-emerged in the national *gaokao* ('higher exam') introduced by Mao in the 1950s, and still, as discussed in the last chapter, going strong today.

Reading about the history of the Chinese examination system today, it's hard not to draw parallels with our modern system of elite global education. Both systems caused extreme anxiety among students. Both are plagued by problems with cheating, and are over-reliant on formalistic essays. And today's overworked university lecturers will no doubt sympathise with their Chinese forebears who didn't have enough time to mark their students' work properly.

But perhaps most strikingly, the Chinese system was tangled up with a set of ideas about mental ability and inequality that echo many of those we see around us today. It was a system suffused with a set of assumptions: if you were smart enough and you worked hard you could achieve educational success; and if you achieved educational success you would have access to the jobs that would guarantee you a rewarding, secure and prosperous life.

Society in turn would benefit from your skills and abilities. But these stories overlooked the unequal playing field on which this educational competition played out. And they largely ignored the fate of the majority of people who didn't belong to the educationally successful elite. Just as is the case today, a society supposedly organised around educational and intellectual merit was not necessarily one that promoted equality.

The Enlightenment and the birth of intelligence

To recap, ideas about human reason were developed by the ancient Greek and Arab philosophers, and imperial China sought to find ways, however limited, to draw the most intellectually gifted into government service. But in none of these times and places was 'intelligence' understood in anything like the way it came to be by the start of the 20th century.

Our modern ideas about intelligence really began to emerge during the Enlightenment. A period that celebrated human reason and its power to overcome superstition and fanaticism naturally prompted renewed interest in what such reason consisted of, who possessed it and what role it should play in the organisation of society. The spread of science, education and new ideas about democracy sparked big debates about human mental abilities.

The fact that our modern ideas about intelligence began to emerge during the Enlightenment doesn't mean religion wasn't an important part of the process. Just like the philosophers of the Islamic golden age, Christians had long been interested in human reason and its relationship to God. Following the Reformation, Protestants believed that the authority of the Church could no longer be relied upon for a true understanding of God and His creation. Instead, people needed to apply their own knowledge and reason to scripture, and to understanding the world around them. Puritans and dissenters held that salvation or damnation were predestined. But they also believed that education and understanding were necessary to ensure that the chosen would be worthy of God's grace.[13] Knowledge was a route to salvation, and logic a means of understanding the world God had created. From here it was a short jump to believing that human society should be led by an elite of the saved, whose logic and learning distinguished them from their fellow man.

This Christian vision of humanity – divided between the educated and able on the one hand and the ignorant and incapable on the other – points us towards one of the key ways ideas about intelligence were evolving. The increasingly modern understanding of intelligence that began to develop during the 17th and 18th centuries was never driven merely by an Enlightenment-inspired celebration of human reason. It was as much, if not more, about the desire to establish hierarchies and to exclude and control certain groups.

Just as in earlier periods, there were lots of different ways that people talked about intelligence during the Enlightenment. One term that was increasingly prominent in Europe during the 17th and 18th centuries was 'genius'.

The word had a much longer history, evolving from separate Latin terms referring to an individual's spirit, and to ideas about character and aptitude. From the 18th century, genius became increasingly linked to ideas about creativity, invention and imagination. Philosophers such as Kant sought to define the nature of aesthetic genius and creativity. The way genius was understood gradually changed from being an attribute all people possessed to some degree to one possessed only by a few exceptional individuals. Writers including Victor Hugo and Thomas Carlyle tried to identify the geniuses of the past and to uncover their impact on human art and history. And there was a popular surge of interest in identifying contemporary geniuses, with newspapers and children's books becoming increasingly interested in stories about child prodigies.[14]

The role of intelligence in this new vision of genius was complicated. Intelligence was sometimes understood as one among a number of different elements that went into making up genius. But it could also be understood as its antithesis. Whereas genius was increasingly associated with madness and instability, particularly for the Romantics, reason and intellect were often dismissed as the narrow ability to apply rules, lacking the spark needed for true creativity. 'Beware of people whose pockets are filled with intellect,' warned Diderot, the author of the *Encyclopédie*, '... They have no daemon. They are not sad, gloomy, melancholic or silent.'[15]

The words more commonly used to describe mental abilities in Europe during the early modern period were 'ingenuity' or 'wit'.[16] Both of these terms could encompass the idea of intellectual ability, but, as in earlier periods, were also linked to notions of temperament and character. Beginning in the 16th century, they became increasingly tied up with ideas about social status. Whereas in the past a person's status was linked either entirely to their birth or to non-intellectual qualities such as martial bravery, now ingenuity or wit came to be seen as one of the characteristics that distinguished a 'gentleman' from the masses of his fellow man. These ideas also became associated with the notion of mental speed. Where both ancient and medieval philosophers had distrusted quick thinking, Renaissance medical writers linked quick apprehension to finer brains. Efficient minds came to be increasingly valued in the context of a developing capitalist economy and the Protestant ethic of labour and productivity.[17]

The 18th century saw another lexical shift towards the concept of 'talents' in Britain and North America, or *'l'esprit'* in France. This was a period that saw lots of debates about where talents came from. There was an almost universal agreement that talents were inherent and inherited. The French philosopher Helvétius caused a scandal in the 1770s when he suggested that mental abilities were the result of education and experience rather than innate qualities of the mind. The implication that all minds were created equal was too much to bear for his contemporaries. Diderot denounced Helvétius for his 'absurdities', while Voltaire thundered that 'nothing is demonstrated more false by experience' than the idea that the only difference between human minds was the education they had received.[18] As we'll see in later chapters, these conflicts would re-emerge in the fierce debates over nature versus nurture in the 20th century.

Wherever talents came from, contemporary commentators agreed that they were increasingly important. The French and American revolutions had upended the social order. After the overthrow of hereditary monarchies, leadership and prestige could no longer be based just on birth or title. The Enlightenment revolutionaries, however, certainly didn't want to see the whole notion of social order overturned. Even as they embraced the rhetoric of equality, they remained strict believers in hierarchy.

Mental abilities offered a way to square this circle. In place of an artificial aristocracy of birth, Thomas Jefferson argued in 1813, society should be governed by a 'natural aristocracy' based on 'virtue and talent'.[19] Everyone would have the opportunity to lead, and no one would be excluded because of their background. But in a world where it was understood that certain people (rich, White, Christian men) possessed greater talents than others (women, the working classes, non-Whites, etc.), equality of opportunity could be promoted safe in the belief that it was unlikely to lead to equality of outcome. Jefferson's idea of a natural aristocracy of talent thus echoed the examination-based meritocracy of imperial China – a system theoretically open to all, but in practice designed to shore up social order by reinforcing elites and excluding most ordinary people.

The belief that talents, reason and intelligence were innate, and that certain groups possessed more of them than others, reflected a broader interest in biological science and classification during the period. Over the course of the 17th century the Swedish naturalist Carl Linnaeus had published his epochal study of the natural world, *Systema Naturae*, which for the first time sought to classify all animals, plants and minerals, including humans and human races. Linnaeus's work reflected the Enlightenment goal of ordering the natural world under the purview of human reason. It inspired a new generation of scientists and social commentators to pursue theories about human difference, hierarchy and classification. And in order to uphold the theory that certain groups were naturally distinguished by their superior intelligence, Enlightenment thinkers needed evidence of a less intelligent 'other' over which they could claim the right to rule.

One such group included those who might today be labelled as people with learning disabilities. Again, there's a long and complex history of ideas about such groups, but the 17th and 18th century witnessed new systematic attempts to define and categorise them. The English philosopher John Locke, for example, argued that humans were defined by their understanding and capacity for abstract reasoning, but saw 'natural fools' as the exception.[20]

In England, interest in such groups was also driven by the demands of the legal system. If someone lacked the intelligence needed to manage their own affairs, then they weren't allowed to bequeath their own property, enter into legally binding contracts or give testimony in court. 'Idiots' had been a legally defined group since the medieval period, and those identified as such became wards of the monarch. In the 17th century the term 'imbecile' was introduced to identify those who were perceived to lack normal intelligence but who were not categorised as 'idiots'.* To help courts accurately identify such individuals, tests were developed to assess a person's literacy and numeracy, their knowledge of themselves and others, and their ability to grasp abstractions and ideas.

These ideas about 'idiots' and 'imbeciles' also became important when Europeans began thinking about the intelligence of non-European peoples. As European travellers, colonial administrators and settlers came into contact with people of other races, they often struggled to understand what was going on in their heads. This was particularly the case when Europeans weren't met with the fascinated amazement they thought they deserved from non-European peoples, or when they observed behaviour that they would have understood as personally or sexually immoral at home. Colonial travel writing, a vastly popular genre, was full of stories about the stupidity and irrationality of non-Europeans. These descriptions often highlighted the same behaviours (innocence, laziness, emotionality) and physical characteristics (facial expressions and features) that characterised accounts of 'idiots' and 'imbeciles' at home.[21]

---

* The history of intelligence is full of offensive, discriminatory and downright hate-filled language, many examples of which, in order to help us fully understand this history, are reproduced in the book.

The comparisons between colonial subjects and European 'idiots' bled into new legal theories about race. The papacy had already given free rein to Spain and Portugal to colonise and enslave non-Europeans from the 15th century. In the 1620s, the Dutch legal scholar Hugo Grotius provided a theoretical scaffolding to this system by arguing that non-European peoples weren't subject to international law or to laws of property and commerce because they lacked reason – the same argument that was used to restrict the rights of 'idiots' at home.[22] In the same way that European 'idiots' needed a guardian – whether the king or a responsible family member – to look after their interests, so the irrational indigenous peoples, Grotius argued, needed European guardians to take care of their lands and their interests.

The scientists who followed in the footsteps of Linnaeus were also formulating new theories about race and intelligence. Christian thinkers had long debated the origins of the various human races, including arguments in the 17th century that indigenous peoples were so-called 'pre-Adamites', descendants of a separate human species who had lived before Adam.[23] Ideas such as these helped to inspire new scientific attempts to understand racial differences that emerged from the 18th century.

At the heart of these efforts was the idea that non-White races were mentally inferior, an idea advanced in early works of comparative anthropology, in medical studies of racial difference, and in the increasingly popular study of the size and shape of human skulls. Many European thinkers regarded Africans and the indigenous peoples of Australia and the Americas as imperfectly rational creatures. Following Linnaeus, they sought to place them within the hierarchy of human and animal species. If man was first among the primates, then mankind itself could be categorised in a hierarchical way. White, rational Europeans were unambiguously at the top of this hierarchy. Other races (alongside the 'idiots' and 'imbeciles' of the European races) could be ranked beneath them, standing somewhere between European man and the most advanced of the animal primates.

These ideas were perhaps most clearly expressed in one of the foundational texts of modern scientific racism, the *Essay on the Inequality of Human Races*, published in 1853 by the French writer Arthur de Gobineau. Gobineau argued that mankind was divided into different 'parts' or 'categories', and that the inequality between them stemmed from differences in intellect. Such intellectual differences were reflected in the levels of civilisation different races had achieved, and the difficulties European colonisers had in 'communicating' civilisation to the various peoples they ruled.

Gobineau's views about the intellectual inferiority of other races were linked to his beliefs about intellectual differences among different classes of White Europeans. He was keen to make clear that not all Europeans were intellectually superior to all non-Europeans. To advance such an argument, he said, would be to claim 'that every European is intelligent; and heaven keep me from such a paradox!' Indeed, he argued that many 'negro chiefs' were the intellectual superiors of most European peasants, or the 'average specimens of our half-educated middle class'.[24] Gobineau's intelligence-based racism was tied up with his intellectual disdain for Europeans outside of the aristocratic elite.

These arguments about race and intelligence were turbocharged by new theories about human evolution. In his 1871 work, *The Descent of Man*, Darwin argued that human mental abilities did not represent a completely different class of intelligence from those of animals, but only differed by degree. Although Darwin himself had little interest in studies of racial difference, this line of thought encouraged the notion of a continuous hierarchy of intelligence stretching up through the animal kingdom and the higher primates into gradations of human intelligence, with rational Europeans at its pinnacle. Herbert Spencer – the man commonly credited as the founding mover of Social Darwinism, which applied the theory of evolution and ideas about the survival of the fittest to human society – was among the first to set out these notions systematically. Calling for a 'comparative psychology' to investigate mental differences between the human races, he described what he saw as the differences between the raw intelligence, mental complexity, mental flexibility and rates of cognitive development among different races, as well as the differences in particular skills and character traits. The mental abilities of races, he believed, could be compared and contrasted just like the intelligence of humans and animals, or of adults and infants.[25]

Not everyone shared these views. Those who sought to challenge the new 'scientific' racism often did so by contesting such claims about intelligence. In the 1780s, for example, abolitionists were on the lookout for stories of enslaved Africans with obviously high intellectual faculties which could be used to counter claims about low Black intelligence. One case they unearthed was that of Tom Fuller, an African man who had been enslaved at the age of 14 and brought to the US, where he had spent his life working on a farm in Virginia. 'Negro Tom', as he was called in the press, was a mathematical prodigy, able to rapidly calculate in his head the number of seconds in any given number of years, or the number of yards in any given distance. He never received a formal education, and opponents of slavery lamented that this neglect had potentially deprived civilisation of another Isaac Newton.[26] At the same time as Fuller's story was coming to light, the famous abolitionist and formerly enslaved person Olaudah Equiano was explaining to his British readers that 'understanding is not confined to feature or colour', and the reasons that were commonly given for labelling Black men and women as unintelligent were in fact the results of slavery and oppression.[27]

In the 1850s, the African-American abolitionist Frederick Douglass advanced similar arguments against those who dismissed evidence of Black intelligence, denouncing the scientific theories of his day 'that would connect men with monkeys; that would have the world believe that humanity . . . is a sort of sliding scale, making one extreme brother to the ourang-ou-tang, and the other to angels, and all the rest intermediates!'[28] A few decades later, Haitian scholar and Pan-Africanist pioneer Anténor Firmin took specific aim at the pseudo-science of Arthur de Gobineau. In *The Equality of the Human Races*, Firmin systematically dismantled the attempts by anthropologists, psychologists and other White scholars to prove the existence of an intellectual hierarchy among human races, whether by measuring skulls or by weighing brains. He questioned how anyone could make claims about Black intellectual inferiority given the extraordinary extent of mathematical knowledge in ancient Egypt, or the cultural or intellectual achievements of Haitians since they had won their freedom a century earlier. The fact that all these attempts to prove the intellectual inferiority of certain races were so riddled with empirical and logical errors just went to show, he believed, that they were 'totally alien to science'.[29]

These debates about race and intelligence were mirrored in debates about the rights of working people and of women. Early modern writers and thinkers such as Samuel Johnson tended to associate common people with dull wits. Those who wanted to extend the suffrage had to argue that the intelligence of the working classes wasn't naturally inferior to that of their social betters. Women also tended to be seen as 'natural idiots', incapable of abstract reasoning. Medical writers linked this to the smaller size of their brains, and women with speed of wit were painted as suspicious and unstable.[30] Mary Wollstonecraft, in her efforts to bring the rights of women to the attention of her peers, had to argue that this 'appearance of weakness' was actually the result of defective education and restricted freedoms, rather than any biological inferiority.[31] Other Victorian and Edwardian feminists challenged the assumption that genius was inherently male, arguing that truly free women would be able to demonstrate the same greatness of mind as men.[32]

The increasing interest in intelligence, then, was tied up with debates about power and inequality. The period between the 16th and 19th centuries saw the emergence of new ideas about political rights, slavery, democracy, empire and freedom. Arguments about intelligence were sometimes used to challenge the oppression of particular groups, to advocate for rights and freedoms. But more often than not, the idea of intelligence served to legitimise inequality, providing a justification for the rule of certain groups over others.

By the second half of the 19th century, the way intelligence was understood had begun to look similar to the ways in which we understand it today. The term itself was increasingly being used to refer to high mental ability. It was becoming more closely associated with reason and rationality, losing its link to ideas about virtue and character, as well as its entanglement with notions of religion and the divine. It was understood as something that certain individuals and groups possessed more than others. And it was widely accepted that these intellectual differences could justify differences in the rights and freedoms particular groups enjoyed. These new ideas were becoming embedded across society, from political debates about education and meritocracy to the novels of eminent Victorian writers.[33]

In the 1850s, Arthur de Gobineau had ended his essay with the hope that a means would one day be found to measure intellectual power 'mathematically'. This, he predicted, would provide scientific evidence for the intellectual superiority of the White race and of elites within Western societies. Just a few years later, the first modern psychologists began to study mental abilities and found, in the new ideas about intelligence, a conceptual toolkit and a convenient label to build their work around. They set about turning intelligence into a science. The scientific racists of the 20th century would eagerly seize on this new science to try to do just what Gobineau had hoped for.

3

The Birth of Intelligence Science

'The tendency has always been strong to believe that whatever receives a name must be an entity of being, having an independent existence of its own. And if no real entity answering to the name could be found, men did not for that reason suppose that none existed, but imagined that it was something peculiarly abstruse and mysterious.'

John Stuart Mill

BY THE MID-19TH CENTURY the way intelligence was understood was still in flux – increasingly defined as the mental power of a person or group, but retaining some of its older links to notions of information, virtue or the divine.

At the same time as these changes were taking place the new science of psychology was emerging, a science that sought to understand and measure the workings of the mind. Up until this point, the study of the mind had been the purview of philosophers. Indeed, many of the earliest psychologists held university positions in philosophy.

But a new generation of men, such as Wilhelm Wundt in Germany, sought to wrench the mind from the grasp of the philosophers and to subject it to modern scientific methods. Psychology, Wundt believed, was something that should be carried out in laboratories rather than seminar rooms, built upon practical experiments rather than airy philosophising.

Intelligence was not initially a topic of much interest for these early experimental psychologists. When they considered the issue at all, they tended to believe that cognitive ability consisted of lots of different, independent faculties. One could study, they believed, how good a person was at calculating numbers or understanding language. But they saw these as fundamentally separate faculties and didn't assume there was a single power underpinning them that could be subject to experimentation.

All that was to change in the short period at the end of the 19th century and the start of the 20th. During that time, a group of psychological pioneers in Britain, France and the United States developed new ways to study and measure human intelligence. In doing so they transformed it from an ill-defined, disparate idea into something that appeared to be concrete, unified and extremely important. Rather than treating intelligence as a common characteristic of humankind, they focused on how much it differed between individuals and groups — something some people had a lot of and others only a little. Excited by their new discovery, they offered it up to their respective societies as a new scientific tool to categorise and manage populations.

And in doing so they embedded intelligence into the heart of another new scientific field that was emerging during the same period: eugenics.

Galton: measuring minds

One of the first people to try to scientifically study human intelligence was Francis Galton.

Galton was the very model of an eminent Victorian. Born in 1822, his family was both rich and intellectually prestigious. His grandfather was the famous doctor and natural philosopher Erasmus Darwin and Charles Darwin was a cousin. This background bequeathed to Galton both an inclination towards science and the financial freedom to pursue a career as a 'gentleman scientist'.

Galton originally studied medicine before switching to mathematics at Cambridge. His early professional life was somewhat directionless, but he found his initial fame as an explorer (another archetypal Victorian profession), leading expeditions into South West Africa (modern-day Namibia) on behalf of the Royal Geographical Society, and then promoting his experiences through popular travel writing. He then turned more seriously to science, and over the following decades pursued an eclectic research agenda across what we now understand as completely different disciplines, from meteorology to statistics. Among his achievements were the discovery of the anticyclone; the statistics of correlation, regression and normal distribution; the science of fingerprinting; and the study of twins as a way to understand human genetics.[1]

Galton was particularly interested in the scientific measurement of people, and their mental and physical abilities. During the International Health Exhibition held in South Kensington in 1884 he set up an 'anthropometric laboratory' to collect the biostatistics of members of the public. Visitors entering the laboratory went through a series of tests that measured and recorded characteristics like height and eye colour, physical abilities like speed and strength, and mental abilities like reaction times and visual judgement. Like many public school- and Oxbridge-educated men of his era, Galton saw physical and mental strength as two sides of the same coin and wanted to test how they were correlated within the human population. Although the earliest experimental psychologists had already begun to develop tests like these in Germany and other countries, this was the first large-scale attempt to scientifically measure and assess human capabilities.

Galton's major contribution to the scientific study of intelligence came in 1869 with the publication of a new book, *Hereditary Genius*. The book promised to uncover the secrets of 'genius' through statistical analysis of 1,000 eminent British men (no women, because of 'decorum', apparently) from 300 families, including the leading statesmen, military commanders, judges, writers and scientists of recent history. Galton later regretted the choice of the word 'genius' for the title, arguing that he was actually studying the kind of exceptionally high abilities possessed by around one in every 4,000 people. These exceptional men, he concluded, were almost always

Galton's anthropometric lab (1885)

members of eminent families, with ancestors or descendants who enjoyed similar status and success. According to Galton, his study proved decisively that the mental abilities on which the eminence of these men rested was inherited, that it was a product of nature rather than nurture.

Now, an eagle-eyed reader may wonder whether this particular research agenda was in some way related to the fact that Galton himself came from an 'eminent' family and that he clearly had a high opinion of his own intelligence. Indeed, he was fairly open about the fact that his study had been inspired by his own life and experiences.

Growing up, he said, he had been struck by how many of the most successful people he met at school, university and beyond were themselves descended from illustrious men. And the former Cambridge mathematics student chose to root his study of genius in a detailed analysis of the achievements and pedigree of . . . Cambridge mathematics students.

These life experiences had given Galton a very particular view about intelligence and its significance. Life, he argued, is like a constant exam, and those who succeed socially and professionally do so as a result of their intellectual merits. He had absolutely no truck with the idea that children were born with roughly the same abilities and that only effort or character distinguished them. 'The experiences of the nursey, the school, the University, and of professional careers,' he argued, offered a chain of proofs to the contrary.[2] He accepted that good educational and social conditions could boost mental powers, but believed that the achievements of any single individual had hard biological limits. The mental grasp of most people, he lamented, was 'ludicrously small', and the difference between the highest and lowest intellects was 'enormous'.[3] He accepted that intelligence wasn't the be-all and end-all of success in life, but thought that anyone with a combination of mental ability, 'zeal' and a capacity for hard work would inevitably rise to the top of society.

But there was more to Galton's vision of intelligence than the life trajectory of an eminent, over-educated Victorian. His research was also driven by his ideas about race and eugenics.

The importance of race in Galton's thinking was made clear right at the start of *Hereditary Genius*. The idea for his research, Galton explained, had emerged during a previous enquiry into 'the mental peculiarities of different races'.[4] Later in the book he went on to compare the abilities between what he labelled the 'negro' and 'Anglo-Saxon' races. Drawing on his own experiences of 'exploration' in Africa, he claimed that 'it is seldom that we hear of a white traveller meeting with a black chief whom he feels to be the better man'.[5]

The importance of eugenics was equally explicit. 'Eugenics' was a term Galton himself coined in 1882 to describe the idea of improving human genetic stock, and he was the pioneer of a movement which, after his death, would become one of the most influential and dangerous of the 20th century. As a result, Galton's name has become synonymous with the development of racial science. In 2020, University College London, where Galton endowed a professorial Chair of Eugenics, was forced to remove his name from lecture theatres and buildings.

Hereditary Genius was published over a decade before Galton came up with the term, but eugenic ideas were already embedded in the book. If mental abilities were, like physical features, largely hereditary, Galton argued that it would be possible to breed a 'highly-gifted race of men' just like we breed horses or dogs with particular abilities. The progress of civilisation, he believed, had been accompanied by great strides in human intelligence. And 'if we would raise the average standard of our race only one grade,' he enthused, 'what vast changes would be produced!'⁶ But he also warned darkly that current social trends, particularly for richer women to marry later and have fewer children, were actually leading to the 'degradation' of the human genetic stock. As we'll see later in the book, this was an idea that recurred in debates about race and intelligence throughout the 20th century.

Galton, then, had taken the first steps in turning intelligence into a science, and had grounded his studies in the idea of inequality between individuals and races. His ideas had an immediate impact on the way intelligence was understood in late-Victorian Britain. When the first edition of *Hereditary Genius* was published in 1869 it received a lukewarm response from reviewers, many of whom questioned the treatment of 'genius' and thought that Galton had downplayed the way people's abilities were shaped by the environments they grew up in. But by the time the second edition was published, in 1892, the reception was much more positive. Now no one seemed to have any issue with the idea, either that genius could be measured, or that it was inherited. Something had clearly changed in the last few decades of the 19th century. The world, it seemed, was ready to think about intelligence in a new way.

Binet: testing children
The person who did more than any other to shape the way we came to think about intelligence in the 20th century was Alfred Binet.

Binet was born in Nice in the south of France – then part of the Kingdom of Piedmont-Sardinia – in 1857.[7] Although he hailed from a long line of physicians his medical studies were short-lived (partly because he disliked dealing with dead bodies), as was an abortive legal career. Eventually he landed on the emerging field of psychology and, almost entirely self-taught, he went on to become a pioneer of the discipline. His major achievement, and the thing for which he is remembered today, was the invention of the intelligence test.

Binet could be an aloof and reserved figure, and his distant relationship with his colleagues partly explains the lack of professional recognition he received in France during his lifetime. But he was much warmer with his friends and family, and was particularly close to his two daughters. He largely educated them himself at home, and he seems to have been an attentive and loving father, writing comic plays for his children to perform and taking them out to enjoy the newfangled hobby of bicycle riding.

Binet's daughters played a crucial role in the development of his research. Like many parents, he was struck by the fact that his children, although raised in similar ways in the same environment, had developed strikingly different personalities. Madeleine, his eldest, was introverted and self-critical, but precise, quick and serious when faced with tasks. Alice, his youngest, was more extroverted and imaginative, but lacked her sister's ability to concentrate. Unlike most other parents, however, Binet was dedicated to experimental psychology and fascinated by the scientific study of individual differences, so he decided to turn these observations about the different personalities of his children into a decade-long experiment. He developed a battery of 20 tests designed to gauge the personalities of his daughters, from word associations and descriptions of images, to memory tasks and tests of reaction times.[8]

These tests formed the basis of his early research. Although primarily evaluations of personality and character rather than of intelligence, they offered a model for how experimental psychologists could begin to pin down and measure our hazy notions of mental faculties.

There were two other paths that led Binet from the study of his daughters' personalities to his development of intelligence tests. The first path involved his experiments with psychiatric patients at the famous Salpêtrière hospital in Paris, in which he explored topics such as hypnotism and the unconscious. The second was his interest in children, education and pedagogy, including the issue of 'mental fatigue' in schools, and the study of children with learning difficulties. Around the turn of the century, he began to play a leading role in the Society for the Psychological Study of the Child, which brought together French psychologists, teachers and educational administrators.

In 1904 the society sent a proposal to the French Ministry of Public Instruction asking them to establish a 'Ministerial Commission for the Abnormal'. The 'abnormal' in question were children with learning disabilities. In 1881 the French government had introduced compulsory universal primary education, and since then teachers and school leaders had been trying to work out how to accommodate such children into the school system. Although other countries had begun to develop special classes and schools in the 1890s, nothing similar existed in France. For Binet, who was appointed to the commission, the first and most important task was to devise a simple, objective test that could be used to accurately identify and classify such children and ensure that they, and only they, would be allocated to the new special schools.

Binet set about creating such a test alongside his collaborator, Théodore Simon. In 1905 they published the first Binet-Simon tests, entitled 'New methods for the diagnosis of the intellectual level of the abnormal'. Some of the tests were similar to those he had given to his daughters, combined with a range of others designed to measure a child's judgement. Although Binet had long sought simple physical tests that would indicate intellectual ability – things like head size or handwriting – he had long since concluded that a large range of tests were needed, because there would never be a single test that succeeded in measuring something as complex as intelligence.

The most important innovation that Binet and Simon introduced was the idea of an age-specific mental level. The two men came up with a hierarchy of intellectual tasks ordered by difficulty, and tested them on groups of school pupils to work out at what age 80–90 per cent of children could carry them out. The scale didn't set up a fixed standard of intelligence which children either met or didn't, but showed at each age what level the majority of children would be expected to reach. This, they believed, could then be used to identify those who had fallen behind.

In language that sounds deeply offensive today but reflected the labels of the time, such children could then be more precisely categorised as 'idiots' (lowest abilities), 'imbeciles' (intermediate abilities) or 'feeble-minded' (*débile* in French, the closest to 'normal' abilities). The first American translators of the scale were reluctant to translate *débile* as 'feeble-minded' because the phrase was commonly used in English as a catch-all term for all people with low intelligence. Instead they decided to coin a brand-new word based on the Greek word for 'foolish'. Thus was born one of the most popular modern insults in the English language: moron.[9]

The 1905 tests didn't make much of a splash, but the first revision, published in 1908, was much more successful. Binet and Simon described the new version, more ambitiously than they had done previously, as a way to measure 'the development of intelligence in the child'. This version introduced new tests designed to better measure the higher mental faculties, and was standardised against a much bigger group of children. Crucially, it also provided a scale against which a child's intelligence could be measured. Although it remained largely overlooked in France, the 1908 scale was widely adopted abroad, and most widely of all in the United States. By 1916, it had been used in over 11 countries, and 88,000 testing forms had been distributed in the US alone.[10] Supporters were particularly attracted by the idea that a child's failure at school no longer needed to attract the moral blame it had previously, but could now be scientifically studied, understood and potentially ameliorated with special support.

Binet and Simon were reluctant to offer a fixed definition of intelligence, partly because they saw it as consisting of lots of different faculties. The definitions they did offer tended to focus on quality of judgement and a person's ability to act in a way best adapted to their environment. Their tests were designed to measure the various abilities that made up these qualities. Certain questions were designed to test fairly straightforward faculties, like memory, comprehension and spatial reasoning. Some were more complex, like having to explain the absurdity of sentences such as 'I have three brothers, Paul, Ernest and me'. But others seemed to rely much more on understanding the social and cultural norms of bourgeois men in late 19th-century France. One question, for example, asked what you should do if a person has offended you and comes to apologise. Another was intended to measure the ability that Binet and Simon called 'aesthetic comparison', arguing that a 'sense of the beautiful' was a constituent part of intelligence. They decided that the best way to test this was to draw three pairs of women, some 'pretty', some 'ugly' and some 'deformed', and ask children which in each pair was prettier.[11]

Their vision of intelligence shouldn't be confused with the concept of IQ, an idea which was invented after Binet's death and which Simon explicitly rejected. Binet certainly thought intelligence was important, both for the fate of individuals and for the health of society. But his idea of 'mental level' did not suggest a singular, fixed measure of

Taken from Binet and Simon, The Development of Intelligence in Children *(1916)*

overall intelligence. He was sceptical about arguments from people like Galton that genius was hereditary, and believed firmly that intelligence was 'educatable'. He also differed from Galton in his ideas about intelligence and life success. Whereas Galton had seen life as a constant exam where the most intelligent would naturally win out, Binet saw it as a 'combat of characters' more than a 'conflict of intelligences'.[12] The qualities of will, attention and courage were needed if intelligence was to translate into worldly success.

Spearman and Terman: inventing intelligence

Binet's tests were an important step along the road towards the foundation of a new science of intelligence. But the relatively nuanced way he discussed the nature of intelligence, its origins and its significance, was still a long way from the more muscular and self-confident language of IQ that would emerge by the 1920s. Binet's tests had turbocharged research on intelligence and testing among psychologists, particularly in the United States, and two men in particular helped to transform them over the following decades: Charles Spearman and Lewis Terman.

Spearman was an English psychologist and statistician who worked at Galton's alma mater, University College London. He was fascinated by the work of people like Galton and Binet, but felt that no one had yet got to grips with what the new intelligence tests were actually measuring. Where Binet had been content to average out the scores from lots of different tests to identify vaguely defined 'mental levels', Spearman argued in 1904 that underlying all of these different measures of intelligence was a real existing thing called 'general intelligence'.[13]

Spearman spent the following decades refining this argument. One way he did so was to move away from the term 'intelligence' – which he thought was so ill-defined in common usage as to be effectively meaningless – and to refer instead to g. Alongside the general factor of g, he also posited the existence of specific factors, which he labelled s. All abilities, he argued, involved some degree of g, which individuals possessed to a greater or lesser degree. But in any given ability this *g* was always combined with some *s*, which was specific to that ability and which individuals possessed in different degrees. Someone who was excellent at playing the piano, for example, would be demonstrating a combination of both their general intellectual ability and an ability specific to the field of music. Measuring someone's *g*, Spearman argued, 'will tell us nearly everything about some of his abilities and something about nearly all of them'.[14]

Spearman's theory was significant because it filled the vacuum left by Binet's refusal to offer an unambiguous definition of the intelligence his tests were meant to be measuring. It provided a theoretical foundation for those who understood intelligence as a single power that varied between individuals.[15] And, despite the considerable debates his ideas provoked among psychologists during the 1920s, it was ultimately the theory that became embedded in the new generation of intelligence tests developed in the United States by Lewis Terman.

Terman was the person who introduced the world to IQ and turned intelligence testing into a mass phenomenon. Born in 1877 in Indiana, he spent his early career working as a high school teacher and principal, before turning to psychology and establishing himself at Stanford University in California. Like Binet, he was interested in education, and counted himself among the progressive reformers looking for ways to bring the American education system into the 20th century. But he didn't share Binet's interest in the problems of children who struggled in normal schooling. Instead, he followed Galton's lead in pioneering the psychological study of the highly intelligent. In the 1920s he would launch the first longitudinal study of gifted children, which helped to inspire the gifted education movement that we'll learn more about later in the book.

But the work Terman is best known for today is the translation and revision of Binet's tests, which he first published in 1916, and which was known as the *Stanford Revision of the Binet-Simon Scale*. Inspired by Binet's tests and the excitement they'd sparked among American psychologists, Terman had begun to carry out his own tests in schools near Stanford in 1910. His 1916 revision built on this experience by expanding Binet's original set of tests and extending the age range so they could be taken by adults. And, inspired by a proposal from the German psychologist William Stern a few years earlier, he replaced Binet's language of 'mental levels' with the new concept of 'intelligence quotient', or 'IQ'– the ratio of mental age (as determined by the tests) to chronological age, times 100.

For the first time in history, therefore, it was now possible to give someone a test that would produce a single number offering a definitive assessment of their intelligence. This was Spearman's *g* brought to life. It was also, as far as Terman was concerned, an indication of someone's native, inherited faculties, not their educational level or socioeconomic background. Like Galton, he believed that intelligence was hereditary.

But as with Binet's test, the foundation these claims were based on didn't seem entirely objective. Terman had standardised his revision against a larger cohort than had Binet and Simon, but it was a cohort drawn mainly from middle-class, White-majority schools near Stanford. Many of the questions it set also clearly required knowledge of cultural norms and social wisdom identified with White, bourgeois American society. Test subjects were asked, for example, to explain why we should judge a person more by their actions than their words, or to correctly interpret the message of a fable about Hercules. 'An Indian who had come to town for the first time in his life saw a white man riding along the street', read one question. 'As the white man rode by, the Indian said – "The white man is lazy; *he walks sitting down.*" What was the white man on that caused the Indian to say "He walks sitting down".' Examiners were given a form to record details of participants and their background. One of the prompts asked about the 'cultural status of the home', with options ranging from 'very superior' to 'very inferior'.[16]

As we'll see in later chapters, these tests quickly became entangled in controversies over intelligence and race. Terman amended some of his later tests to accommodate those for whom English wasn't a first language. But he was more than happy to classify racial groups by their intelligence, and showed little interest in criticisms about his sampling methods or the cultural loading of his questions. When he was collecting children with IQ scores of over 140 for his gifted children study, he noted that children with English, Scottish and Jewish parentage were overrepresented, and that there were low proportions of

Mexicans, Italians and Blacks, offering this as proof of different native levels of intelligence between different races.[17] Like so many of those involved in the birth of intelligence science, he was also an enthusiastic eugenicist and active participant in the work of the Eugenics Society.[18]

Despite these obvious limitations, Terman's tests proved wildly popular. This was helped by the fact that, just a year after their publication, the United States entered the First World War. Terman was appointed to a panel of psychologists designing a set of intelligence tests to be given to new army recruits. Nearly 2 million American servicemen sat these tests before the end of the war, with results helping to dictate who would be sent for officer training.[19]

The army tests and their results (of which we'll read more in the next chapter) received huge media attention. They seemed to offer an official endorsement for the new, and until then largely untested, technology of intelligence testing. At the end of the war there was an enormous demand for these tests from teachers and school administrators. Pirated copies of Terman's test quickly sprang up across the country. In response, Terman rolled out new versions of the tests that could be administered to groups rather than relying on expensive one-to-one interviews, as well as other tests for scholastic achievement. By 1925 these tests were selling one and a half million copies a year.[20] Terman was inundated with requests to approve translations and revisions from countries across the globe, including Peru, Mexico, Poland, China and India.[21]

The intelligence-testing industry had been born, and was quickly becoming a global phenomenon.

The story I have told here is by no means a complete history of the birth of intelligence science. There were lots of other people involved in the process, and plenty of other ideas about what intelligence was, how it could be measured and what it meant for the way society was run.

For example, eagle-eyed readers will have noticed that – as is so often the case in history books – not many women have so far appeared in our story. This doesn't mean that women weren't involved in the birth of intelligence science. In fact, women played a more prominent role in the early years of psychology than in many other scientific disciplines of the time. This was partly because psychological research often involved working with children or psychiatric patients, which was work that women were seen as naturally suited to. Alfred Binet corresponded with a number of female psychologists working on educational psychology and testing at the start of the 20th century. The first translator of Binet and Simon's test into English was a woman named Elizabeth Kite, who collaborated in some of the earliest American research into intelligence testing. Catherine Cox, who became one of the most prestigious female psychologists of her era, did her PhD with Lewis Terman at Stanford, calculating the IQ of historical geniuses, and collaborated on much of Terman's later work on intelligence.

But there's no escaping the fact that the birth of intelligence science was driven primarily by men, and that's something we should pause to consider. The idea that there were fundamental differences between male and female minds stretched back to the ancient philosophers. Rationality had long been culturally gendered as male, particularly in the West since the Enlightenment – it was something that was widely assumed to characterise the male mind, in contrast to the supposed irrational or emotionally driven female mind. The decision of the early intelligence scientists to focus on measurements of logic and reasoning reflected these cultural values. The new scientific concept of intelligence they put out into the world was one that was shaped by a culture that most highly valued a set of mental abilities commonly associated with masculinity.

For Francis Galton and other early intelligence scientists, the logical conclusion to draw from this association was not just that male and female minds were different, but that men possessed more intelligence than women. This idea didn't hold up to scrutiny for long. Lewis Terman's earliest tests showed that, if anything, women tended to score slightly higher than men. In response to this finding, Terman decided to remove questions that showed a marked gender difference in results. This meant that early IQ tests were specifically designed to produce equal IQ scores for men and women.[22] As unlikely as it might seem, this feature of IQ testing helped to undermine the arguments for women's lower intelligence that had been seen as a matter of common sense up until the start of the 20th century.

Unsurprisingly, however, such arguments quickly re-emerged in a new guise. Instead of claiming that average male intelligence was higher than average female intelligence, psychologists and other researchers developed arguments about greater male *variability* in intelligence. Chief among these was Havelock Ellis, the pioneering sexologist and scholar of gender difference. Inspired by the key role given to genetic variation in theories of evolution, Ellis argued that, while men and women shared similar average rates of intelligence and other mental abilities, men's abilities *varied* more. This meant they were likely to appear more frequently at both ends of the distribution curve. Women would tend to cluster around the average level of intelligence, whereas more men would possess either very low or very high intelligence. This theory offered a neat biological explanation for the apparent greater number of eminent men and male geniuses.

This was one of the cases where the relatively high profile of women in the psychological profession offered a platform to challenge such views. Leading female psychologists such as Helen Bradford Thompson and Leta Stetter Hollingworth sought to dismantle the variability hypothesis. They revealed errors in the research, and set out some of the cultural and environmental factors that offered alternative explanations for the results which male researchers were attributing to biological differences. Their own more carefully designed experiments found no significant evidence for greater male variability in intelligence.[23] Despite their efforts, though, the variability theory has lived on both in intelligence research and in everyday notions of intelligence. In 2006, former U.S. Treasury Secretary Larry Summers was forced to resign as Harvard president after he claimed that the greater variability of male intelligence helped to explain why top universities hired so few female professors in science and engineering.

These debates were part of a wider pattern that characterised the birth of intelligence science. The psychologists and educationalists who developed this science didn't begin with a blank slate. They took as their starting point the ideas that were already swirling around the culture of their day, the notions about intelligence that we saw emerging from the Enlightenment and the era of colonial expansion in the last chapter. These notions were bound up with the way the 19th-century world thought about inequality – between men and women, between rich and poor, and between Black and White. These ideas about inequality, in their turn, became embedded in the science of intelligence and its new technologies, solidified and stuck fast like fossils in a cliff face.

This link between ideas about intelligence and inequality would have a profound impact on the history of the 20th century.

4

Mental Engineering

'We only await the Human Engineer who will undertake the work.'

Henry Goddard

IN SEPTEMBER 1942, a young woman named Margaret visited the labour exchange (what we would now call a job centre) in Glasgow. Margaret was working in a clerical job for a shipping firm. But as the Second World War entered its third year she was desperate to follow in the footsteps of her twin sister, who had joined the Auxiliary Territorial Service – the women's branch of the British Army, whose members served in non-combat military roles at home and overseas.

There was a group of women at the labour exchange to apply for war work that day. Margaret had expected to be called into an office to be interviewed in person by a recruiting officer. Instead, she and the other women were taken into a room and told they would be taking a test. This was not the kind of test that was familiar to Margaret from school. It consisted of 60 pages, each containing different patterns. At the bottom of each page was a series of numbered shapes, and the task was to relate those shapes to the patterns above.

No one explained what the test was measuring or how it was relevant to recruitment for war work. But when she had finished, Margaret was handed a slip of paper by one of the officials at the labour exchange. 'The Ministry of Labour,' it said, 'regrets that there are no vacancies in the Women's Forces for which you are suitable.'

Margaret, understandably, was not happy about this. When she got back to her office she told the women she worked with all about it, and everyone agreed it made no sense for her to be rejected on the basis of such a nonsensical test, without even having the chance to talk to a recruiting officer. Some suspected foul play, alleging that only those with the right connections were accepted for the best roles. Others encouraged Margaret to write to her MP to complain and see if there were other ways she could join up.

One of her colleagues had a husband who worked for an organisation called the Institute of Industrial Psychology, which, among other things, designed recruitment tests for businesses and government departments. When she got home from work this friend told her husband about Margaret's story. The test Margaret had taken, he said, was designed to measure general intelligence. From his experience he would have expected it to be used alongside a face-to-face interview and a medical examination. But although this hadn't happened in Margaret's case, it sounded like the test results had shown that she had too little intelligence to be recruited into the army.

Margaret's colleague thought this was nonsense, and told Margaret about what she discovered when she arrived back into work the next day. Margaret was distraught at this news, insulted by the idea that she'd been officially dismissed as unintelligent on the basis of her performance in a single, strange test on a single day. She was desperate to find out what she could do to challenge it, but no one was able to find a way for her to overcome her initial failure. It was particularly galling, she told her friend, because she knew 'heaps of girls who are absolutely stupid, and their absence of intelligence did not keep them out of the Army'.[1] Margaret spent the rest of the war working in the same office, and never got to join up with her sister on active service.

As I've been researching the history of intelligence, it has not been particularly hard to uncover the stories of eminent men like Galton, Binet and Terman. The psychologists, neuroscientists and computer scientists, educationalists and other experts who studied intelligence from the late 19th century generally wrote lots of books and scholarly articles that are easily accessible today. Many have archives of personal papers and correspondence that researchers can consult. They were discussed in newspapers, appeared in the media and now all have their own Wikipedia pages. They have left a trail in the historical record that is relatively easy to follow.

This is not the case for people like Margaret, ordinary people who found themselves at the pointy end of the new intelligence technologies that emerged during the early 20th century. To uncover those stories, I had to look a bit harder. Margaret's story is told in one of the diaries held in the archives of Mass Observation, a pioneering social research organisation that, between the 1930s and the 1960s, collected diaries and questionnaires from thousands of ordinary people across Britain. Among the pages and pages of accounts of everyday life in the archive, we find many stories about people like Margaret who encountered some form of intelligence test.

Some people seem to have embraced the new testing technologies. There are stories in the Mass Observation archive of people arranging 'IQ parties', where friends and family would get together, take an intelligence test and compare their scores. The next time you're at a party and are not enjoying yourself, be thankful, at least, that it's not one of those parties.

For most people, however, the experience of these tests was much more negative. Those who had tests thrust upon them in the course of their work were often thrown into a panic, with little idea of what they were for or how their performance would be judged. The questions or tasks they were faced with often seemed strange, and most people couldn't work out how they were relevant to real life. For many of them, 'failure' in the tests had tangible negative consequences: exclusion from a job (as was the case for Margaret), or the withdrawal of a scholarship or school place for a child. In the aftermath of the tests there was almost never any information available about how exactly a candidate had performed or what their results meant, and very few options to challenge the outcome.

Few of those faced with such a situation seemed to want to challenge the validity of intelligence as a thing of value or a coherent concept – they accepted that intelligence was something important, and that some people had more of it than others. But this only made the psychological impact of failure worse. Not only were they faced with the practical fallout from the tests, they were burdened by the stigma of officially being judged as intellectually inferior.

This was an entirely new historical phenomenon. Intelligence tests, we must remember, were a very recent technology. Prior to the First World War their use outside of very limited medical or research settings was completely unknown. As we've seen, the idea of intelligence as a coherent quality, something important that varied between individuals, had been gaining ground through the 19th century and had been given a renewed boost by the intelligence scientists of the early 20th century. But so far it had remained something confined to the realm of ideas, something that people might have had opinions about, but which had little impact on everyday life.

Now, though, things were different. Intelligence was no longer just an idea. It had become a tool, a thing that could be used to make decisions about ordinary people and their life chances, something that could be used to manage individuals and engineer society.

The idea of using intelligence science to engineer society – 'mental engineering', as it was sometimes described – rose to prominence after the First World War. It was tied up with contemporary debates about the new age of mass society, about democracy and about the role of intellectual elites. But it wasn't confined to the realm of ideas or abstract questions about politics and society. It had a very real impact on everyday life, reshaping the worlds of work and education in ways that touched the lives of millions of people around the world.

The talented tenth

Underpinning these changes was a new set of ideas about mass society and intellectual elites that emerged at the start of the 20th century. As 'ordinary' people began to gain greater influence over social and political life, there were those who argued that a new, highly intelligent elite was needed to maintain order and to drive progress. A modern mass society, they believed, could only work if it was led by a 'talented tenth'.

The idea of the talented tenth was popularised by W.E.B. Du Bois. Du Bois was probably the most well-known African-American intellectual and activist of the early 20th century. Indeed, he is arguably one of the most important political thinkers of the modern era. Born in Massachusetts in 1868, he studied at Fisk University, became the first Black American to gain a PhD from Harvard, and went on to teach at Wilberforce, Philadelphia and Atlanta universities. He was a pioneering sociologist, public intellectual and activist, who would play a pivotal role in the history of Pan-Africanism and the civil rights movements.

Du Bois's politics changed as the century wore on, and later in life he would come to more explicitly embrace socialism and to celebrate the achievements of communist states. But earlier in his career he combined intellectual radicalism with personal attitudes and behaviours that were more conservative. He didn't come from a particularly wealthy background, but had enjoyed educational advantages in his youth. Later in his life he admitted that he had despised the poor Irish and Germans in the town he had grown up in, and had 'annexed the rich and well-to-do as my natural companions'.[2] People who knew him as a young man often noted his aristocratic bearing and reserved manner, and critics sometimes accused him of being snobbish and haughty.[3]

Some of this might help to explain Du Bois's embrace of the idea of a talented tenth, which has remained linked to his name ever since. It was not a phrase Du Bois coined himself, but he made it famous in a chapter he wrote for his 1903 book, *The Negro Problem*. 'From the very first,' he argued, 'it has been the educated and intelligent of the Negro people that have led and elevated the mass.' If that progress was to continue into the 20th century, it would need to be led by exceptional men, the 'aristocracy of talent and character' who would bring culture and progress to the Black masses. What was needed to make that happen, he believed, was to ensure that this talented tenth received the liberal higher education that would engender the 'intelligence, broad sympathy, [and] knowledge of the world', needed for their task.[4]

It's an idea that today seems elitist, and indeed one that Du Bois distanced himself from in later life, when he became increasingly concerned with class stratification and exploitation within the Black community.[5] But it was an idea that was embedded in the culture of the time. In the aftermath of Reconstruction, Black elites were debating how to achieve what was referred to at the time as 'racial uplift'. Since the 1880s, Booker T. Washington had been promoting industrial and vocational education for Black Americans, part of his broader argument that the Black community should accommodate itself to segregated White rule in the South and focus on economic development, rather than fighting for civil rights and political equality.

Du Bois's 'talented tenth' argument was aimed squarely against these kinds of policies. His attack on Washington and on vocational education marked the beginning of a fierce struggle between the two men and their supporters. But to truly understand where Du Bois's ideas were coming from, we have to look beyond the United States and consider the broader debates about intelligence, elites and mass society that were buffeting societies around the world at the time.

As we can see from the quotes just given, part of Du Bois's understanding of the talented tenth was that it was made up of those with the highest intelligence. He didn't believe in the kind of fixed, hereditary notion of intelligence that was being promoted by some intelligence scientists at the time. Instead, he saw intelligence as something that could be cultivated through education. Indeed, he also wrote about the intelligence of Black workers outside of the talented tenth, and the role they would play in achieving Black uplift. But, like many people of the era, particularly in the United States, Du Bois had faith in the idea of intelligence and its significance. He was interested in the IQ scores of young people in his family, and wrote about the IQs of famous Black figures.[6] He understood intelligence as something that some people possessed more of than others, that was concentrated among society's elites, and that helped to advance culture and drive progress.

This faith was rooted in contemporary concerns about mass society. When Du Bois referred to the 'mass' of African Americans outside of the talented tenth, he was tapping into a set of ideas that had entered the heart of political and cultural life from the end of the 19th century, and would come to define the start of the 20th century. The 'masses' were understood, by people across the political spectrum, to have become the driver of modern societies. Some welcomed this change. But others, particularly those like Du Bois who identified themselves with the cultural and intellectual elite, saw it as something to be feared, controlled or resisted.

This concern about the masses reflected a set of major social and political changes that had been taking place over the course of the 19th century. The Industrial Revolution had drawn millions of people into towns and cities, and had driven the creation of vast new factories and industrial centres. Many democracies had extended their franchise to give the vote to working men and, in a few countries, to women. Political parties had been forced to transform themselves to appeal to these new constituencies. The labour and trade union movements were playing an increasingly important role in political life. A growing middle class was driving new forms of consumption, and increasingly leading trends in fashion and culture. These changes, combined with new technologies, were facilitating the growth of mass culture, leisure and media – from vast football crowds to mass-market newspapers.

All these seemed to herald a new era in which the masses would be ascendent. For those who belonged to the masses, this probably seemed like a good thing. But it caused profound consternation for those who identified with the politics and culture of the previous era.

Take Gustave Le Bon, for example. Le Bon was a prominent intellectual in late 19th-century France. He was a polymath and a science populariser, writing about topics from horse training and anthropology to reproduction and atomic energy. He ran an influential debating club in late 19th-century Paris that attracted leading political and cultural figures of the day, and he acted as a senior advisor to various French rulers in the 1900s and 1910s. A vain man and a womaniser, he was also a reactionary *par excellence*. He hated socialism and feminism, and was a big fan of racial purity and segregation.[7]

Le Bon is best remembered for his 1895 study on the psychology of crowds. It was a work that was influenced by the political turmoil of late 19th-century France. Universal male suffrage had been introduced in 1884, but French politics had been buffeted by the rise of popular demagogues, corruption scandals, the rapid growth of the workers' movement and an anarchist terror campaign that culminated with the murder of the French president in 1894. Le Bon saw this as part of a longer pattern of violence and instability stretching back to the French Revolution and the Paris Commune. All of this upheaval, he believed, was a consequence of the rise of the masses. 'The destinies of nations,' he lamented, 'are elaborated at present in the heart of the masses, and no longer in the councils of princes.'[8] Suffrage and the workers' movement had transformed the popular classes into the governing classes.

He saw the modern crowd as the physical manifestation of this phenomenon, the power of the masses brought to life. The modern era, he argued, was the era of crowds. He didn't think this was something that could be undone or avoided, but he argued that the psychology and unconscious actions of the crowd needed to be understood so its power could be politically channelled and controlled. This was an idea that proved extremely popular among the political leaders and wannabe demagogues of the earlier 20th century, and influenced figures from Theodore Roosevelt and Charles de Gaulle to Mussolini and Hitler.

Le Bon's arguments about crowds were rooted in ideas about intelligence. He believed that the crowd possessed a kind of collective mind, distinct from that of the individuals within it. The problem with the collective mind of the crowd, he argued, was its 'extreme mental inferiority'.[9] The crowd was guided almost entirely by unconscious impulses rather than by reason, and what reasoning it was capable of would always be of an inferior sort. This held true regardless of the intellectual level of the individuals within it. Even in assemblies of intelligent experts, he believed, 'it is stupidity and not mother-wit that is accumulated'.[10] And he feared that the rise of the mentally inferior crowd would have terrible consequences. 'Civilizations as yet have only been created and directed by a small intellectual aristocracy,' he wrote.[11] Now that intellectual aristocracy was being usurped by the irrational crowd, Western civilisation would collapse and anarchy would reign.

His ideas tapped into the Enlightenment tradition of justifying social and racial hierarchies on the grounds of intelligence. The psychological characteristics of crowds, he believed, mirrored those observed in 'beings belonging to inferior forms of evolution' – in other words from those he saw as 'inferior' races. To explain the flawed reasoning of crowds, he compared them to the 'Esquimaux' who believed that glass would melt in the mouth just like ice, the 'savage' who believed he would acquire bravery by eating the heart of a courageous foe, or (the very height of irrationality) 'the workman who, having been exploited by one employer of labour, immediately concludes that all employers exploit their men'.[12]

Le Bon's reactionary vision was a world away from Du Bois's project for the cultural development of the Black community in the United States. But both men were tapping into contemporary fears about the rise of the masses, and the need for the talented tenth to guide or control them.

These fears about the masses and debates about the role of elites in the modern age reached their peak on the other side of Pyrenees in early 20th-century Spain. Spanish politics had been plunged into turmoil in 1898, when military defeat to the United States led to the loss of Spanish colonies in Cuba, Puerto Rico and the Philippines, an event that came to be known simply as 'The Disaster'. This sparked a wave of recrimination and self-flagellation among Spanish elites, who were desperate to explain how a formerly proud imperial power had been brought so low so quickly.

This feeling of national despair was exacerbated when parts of the country were plunged into revolutionary tumult in the aftermath of the First World War. Spanish attempts to re-establish an empire in northern Morocco were initially met with humiliating failure, and the parliamentary regime was overthrown by the dictatorship of General Miguel Primo de Rivera in 1923. Although he was inspired by Mussolini's rise in Italy, Primo de Rivera was much less successful at cementing his political movement in power, and after he was forced from office in 1930, a Spanish Republic was declared. The Republic was greeted with hope that it would inaugurate a period of democratic national renewal, but political life was buffeted by fierce political conflicts between right and left, and the country was plunged into civil war following a coup led by Francisco Franco in 1936.

In the midst of all of this turmoil, Spain's liberal intellectuals became deeply concerned about the impact of the masses on modern politics. Chief among them was José Ortega y Gasset, Spain's foremost philosopher and a high-profile public intellectual during the 1920s and 1930s. In a series of newspaper articles and books written during the period, most famously his 1930 book, *The Revolt of the Masses*, Ortega y Gasset charted what he saw as the dangerous consequences of the rise of mass politics. These consequences, he argued, were evident in all aspects of modern social and political life, from political movements refusing to follow their leaders to middle-class women trying to lead the fashion of the aristocracy.[13]

Despite Ortega y Gasset's liberal politics, many of his arguments echoed those of Gustave Le Bon. Like Le Bon, he argued that human societies were naturally aristocratic in that they were always led by a select minority. It was this intellectual minority, he believed, which drove progress and innovation. Now, he lamented, the 'commonplace mind' of the mass had taken over. And, even worse, the mass, like a spoiled child, had no idea of its own limitations or the superiority of its betters, loudly proclaiming its rights in ignorance of the consequences. When the masses refused to accept the leadership of the select minority, civilisation would crumble and chaos would reign. And he believed this refusal, an 'aristophobia' or hatred of one's betters, explained the chaos and degeneration of contemporary Spanish society.[14]

Intelligence was not as important a part of Ortega y Gasset's argument as it was for Le Bon. Like Du Bois, he saw intelligence as just one of the characteristics of elites, and saw moral and leadership qualities as equally important. He also thought elites needed to come from all parts of society, including women, workers, soldiers, artists and engineers. But he did think that those with high intellectual abilities were vital for achieving social progress. Any society that lacked an elite of 'vigorous intelligence', he feared, would gradually come to think with less rigour, to limit its ideas and points of view, to reduce its scientific and technical capabilities. It would end up as a 'dumbed-down, intellectually degenerated race'.[15]

These fears about intellectual degeneration reappeared constantly in debates about mass society. The French intellectual Paul Valéry – the man who led the League of Nation's intellectual co-operation bureau – marvelled at the transformation of human society in the early decades of the 20th century. But he worried that the pace of modern life, the new technologies people had built to make that life easier, and the new era of massive organisations and bureaucracies were overwhelming and deforming the human mind, 'reducing our powers of attention and the capacity of the average human being for mental effort'.[16]

Valéry's pessimism, however, was not universally shared. For every intellectual who worried that modern society was undermining human intelligence, there was an expert ready to show how human intelligence could be used to save modern society.

Intelligence and work
During the early 20th century, intelligence in general – and the highly intelligent in particular – were increasingly being discussed as a defining force in the modern world. Intellectuals, psychologists and educationalists, particularly in the United States, began to notice a change in the way that intelligence was talked about. Whereas 'cleverness' used to be mistrusted, they argued, now it was becoming more common to see intelligence as a virtue, almost a moral duty that individuals owed to their societies.[17]

This was tied up with the increasing tendency to think about society as a machine. The First World War had seen governments mobilising society and industry on a scale never seen before. Social scientists were developing new theories about how modern mass societies worked and how they could be managed. And it was increasingly believed that modern societies *should* be managed, that industry, commerce, education, industrial relations and all manner of other subjects needed to be directed by wise minds and far-sighted governments, rather than being left to the vagaries of chance or the chaotic forces of the market. In the same way that engineers could design modern machines that were increasingly powerful and efficient, so social engineers should be able to find ways to make society run more effectively.

One big part of modern life that people thought could be managed better was the world of work. This was the era that saw the birth of so-called 'scientific management'. Inspired by management theorists like Frederick Taylor, scientific management offered new modern techniques like time and motion studies to make workers more productive. Economic efficiency, according to this world view, would come from using these new techniques to 'persuade' factory workers to carry more pig iron, fit more rivets or grind more ball bearings per shift than they had previously.

But the intelligence advocates saw things slightly differently. For them, economic efficiency wasn't about finding ways to force individuals to work harder. It was about working out how to make better use of the mental powers that existed across society. The way society allocated different people to different jobs, they argued, was deeply inefficient, too often based on family background, fleeting personal preferences or pure chance. As a result, the wrong person often ended up doing the wrong job. That wasn't just bad for those individuals, but also harmed society as a whole, breeding inefficiency and wasting human talents.

Intelligence and intelligence testing seemed to offer a solution to this problem. As Alfred Binet mused, his new techniques for measuring intelligence promised a future 'where every one would work according to his known aptitudes in such a way that no particle of psychic force should be lost for society'.[18] This view was enthusiastically embraced by a man named Henry Goddard, the American psychologist who was the first person to use Binet's tests in the United States. Goddard wrote rapturously about the way society could be transformed by new ideas regarding intelligence. Almost all social problems, he believed, were rooted in the inefficiency caused by the wrong people filling the wrong roles in society. If people at each level of intelligence could be assigned to the correct part in the social system – one that was commensurate with their intellectual capacity – the machine of society would finally begin to function efficiently. Intelligence testing and applied psychology, Goddard argued, offered the tools to carry this out.[19]

This embrace of intelligence science received a huge boost from the use of intelligence testing by the US Army during the First World War. Prior to the war, intelligence testing had been in its infancy and had only been used on a very small scale. When the United States entered the war in April 1917, the psychologists who had been working on intelligence realised that it offered them an unprecedented opportunity to spread the intelligence gospel. To be successful in modern warfare, they believed, the United States needed to follow the example of the German Army by integrating psychological testing into its recruitment and training.

A committee convened by the American Psychological Association recommended in the summer of 1917 that the army use intelligence tests to help manage its mass recruitment process. Their proposal was approved, initially with the intention that the tests would help to weed out those recruits whose intelligence was too low to submit to the instruction and discipline the army required. The committee – which included Henry Goddard and Lewis Terman, along with other leading psychologists and intelligence researchers – spent a month working feverishly to put together a suitable battery of assessments. What they came up with was a series of tests that included Terman's Stanford-Binet scale, with special versions adapted for recruits who were illiterate or for whom English wasn't a first language. Unlike the early Stanford-Binet versions, which were individually administered, these were group tests, designed to be taken by groups of up to 500 recruits at a time and only requiring a few hours to complete and mark.

American soldiers taking IQ test. Camp Lee, Virginia, November 1917. Source: National Archives

The tests were formally adopted by the army at the end of 1917, and the scale of their rollout was vast. Hundreds of people were drafted in to administer the tests, and in the single year between their approval and the end of the war, they were given to almost 2 million recruits. The way they were used in practice went far beyond the original goal of weeding out the unintelligent. Every recruit was given an intelligence rating, from A ('very superior') to E (recommended for discharge or work in a labour battalion). These ratings were used to select officers and to decide who would be assigned to special duties that required specific mental skills. As Lewis Terman argued, for the army to work effectively, 'the work which requires most brains must be given to men with brains . . . Plainly if the army machine is to work smoothly and efficiently it is as important to fit the job to the man as to fit the ammunition to the gun.'[20] Despite some initial scepticism, senior officers quickly embraced the tests as a vital part of their recruitment and management processes.[21]

In the blink of an eye, intelligence tests had been transformed from a new, niche psychological technology into a mass tool of personnel management and social engineering. Intelligence scientists were quick to grasp this new opportunity. As soon as the war was over, they began to publicise the testing system and to trumpet it as the solution to all manner of post-war problems. Robert Yerkes, the head of the American Psychological Association who had helped design and manage the system, published a detailed series of studies about the tests and their results, which included practical guides for how to administer the tests and copies of blank forms for would-be testers. American success in the war, he claimed, was due to the 'proper utilization of man power, and more particularly of mind or brain power'. The lessons of the war, he argued, now needed to be carried forward into civilian life, using these new psychological techniques as tools of 'mental engineering' to help society run more effectively.[22] He urged school districts to use the tests to give all children an intelligence rating, which could help them decide on the type and duration of education each child should be granted. And he believed that the army's experiences of assessing the intelligence needed for different roles and assigning men accordingly should be applied to civilian industry.[23]

Civilian industry seemed to agree with this. The idea of vocational testing – using psychological tests to fit people to the most suitable roles – had started to develop in the US prior to the First World War. Psychologists such as Hugo Münsterberg at Harvard had developed tests of things like reaction times to help select streetcar drivers and switchboard operators. During the war, the psychologists who had been working on the army intelligence tests had also developed tests to identify particular skills needed for specific roles, such as motor co-ordination to select pilots, or listening tests to select telegraphers.

These kinds of tests expanded rapidly after the war ended. By the early 1920s, factory workers were being presented with tests for measuring mechanical abilities, and office workers with tests to measure clerical aptitude.

A new industry sprang up to publish such tests and to advise companies on how to use them, often with the input of psychologists and intelligence scientists who had realised there was big money to be made from their new techniques. By the 1930s, IBM had even developed special machines to automatically score vocational and intelligence tests on a mass scale.[24]

This new testing craze wasn't just confined to the United States. In Britain as well, the idea that intelligence and other psychological tests should be used to fit the right people to the right job took off after the First World War. The British Army had made some use of psychological testing during the war itself, though on nothing like the scale seen in the United States. But one of the psychologists who had worked with the army, Charles Myers, paid close attention to the American example. After the war he launched his own crusade to convince the British public that mental testing and vocational guidance could transform the productivity and efficiency of British industry.[25]

In 1921, Myers began to put these ideas into practice through a new organisation called the National Institute for Industrial Psychology (NIIP). 'Industrial psychology' was a new idea, roughly equivalent to what we might call occupational psychology today, something which aimed to make organisations more efficient by applying the techniques of modern psychology. Over the following decades the NIIP would play a key role in promoting intelligence testing across the UK, developing a battery of group intelligence tests and training hundreds of testers in how to run the Stanford-Binet tests.

One of the NIIP's most important strands of work was vocational guidance. Although elite schools had long offered their pupils advice about what careers to pursue, there was no national system of vocational guidance or consensus about what it should involve. For people like Charles Myers, effective vocational guidance was vital to the task of transforming modern industry – identifying the talents and abilities of each young person, and then linking them to the job best suited to those talents.

The organisation's work on vocational guidance began in the early 1920s. Psychologists such as Charles Spearman (the inventor of 'g') and Cyril Burt (the pioneering figure in British educational psychology) were drafted in to develop tests and techniques, and the NIIP began offering guidance services from its London headquarters.[26] For a small fee, parents could send their teenage children to undergo a battery of tests and interviews with a psychologist, which would result in a report detailing their abilities and limitations, alongside recommendations about what careers would be most suitable. If, like me, you grew up in the UK and at some point during your school career were given a test which told you what jobs you could do when you grew up, this is where the idea for those tests came from.

The brochure the NIIP produced to promote this service to parents was entitled 'How Bright Is Your Child ... ?', which was a good illustration of how important the idea of intelligence was to the institute's vocational guidance. Although applicants were tested on a range of abilities – including mechanical ability, manual dexterity and temperament – the institute was clear that intellectual capacity was the 'most important' determinant of what job a person should do.[27] Young people were encouraged to ask themselves whether they compared favourably in 'brains' to other people of their age, whether they were quick to grasp an argument or saw themselves as a 'slow thinker'.[28] Vocational guidance, according to this vision, was a tool for assigning people to the correct jobs for their level of intelligence.

All of these ideas about integrating intelligence into the world of work rested on the idea that doing so would increase productivity, efficiency and economic growth. But there was another benefit the intelligence advocates thought these new technologies would bring: the solution to class conflict.

The years following the First World War witnessed intense class conflict across Europe and the United States, with the economic dislocation of the war's end and the example of the Russian Revolution boosting trade union membership and unleashing a wave of strikes and protests. The psychologists and intelligence scientists we've been discussing, firmly rooted as they were within bourgeois society, looked askance at these developments. For them, strikes were a manifestation of irrationality, a symptom of the dysfunctional organisation of modern industry.

Bringing psychology and intelligence into the world of work, they believed, offered a solution to these problems. It would provide a scientific, objective way to organise the workplace, countering extremist tendencies on both sides. Employers would adopt more conciliatory attitudes to workers who had been specially selected to match the requirements of their roles. Workers would respect their bosses if they understood that those bosses had been scientifically selected on the basis of their abilities. And crucially, they believed, matching the right workers to the right jobs would remove the psychological causes of workplace strike. No longer would the 'misfit', placed into a job he was unsuited for, vent his frustration by complaining about his employer and spreading discontent throughout his workplace. 'A man who is doing work that is well within the capacity of his intelligence and yet that calls forth all his ability,' as Henry Goddard argued, 'is apt to be happy and contented and it is very difficult to disturb any such person by any kind of agitation.'[29]

All of this, of course, was nonsense. The idea that workers would suddenly abandon their fight for decent pay because their bosses had been scientifically selected for their high intelligence was pure technocratic fantasy. But it was revealing of the mindset among the psychologists and other experts who had cast themselves as the mental engineers of modern society. As absurd or dystopian as it seems, they genuinely believed that the new intelligence technologies would help make society, and all of the individuals within it, function like a well-oiled machine.

Intelligence and education

The other area in which intelligence was increasingly being used to manage society was education. The early 20th century witnessed a global boom in education. Countries, mainly in Europe and North America, that had already introduced compulsory elementary education in the 19th century now embarked on a big expansion of secondary schooling. Elsewhere in the world, mass elementary education was being introduced for the first time, alongside secondary and advanced education for elite minorities. Education was increasingly seen as the key to prosperity, progress and social cohesion.

But this expansion of education raised questions. Who and what was education for? Should everyone receive the same education, or should it be tailored to the individual? How should decisions be made about who was educated and for how long?

This was where the idea of intelligence came in. Just like with US Army recruitment and the use of testing in the workplace, the modern notion of intelligence seemed to offer an answer to these questions. Where resources were scarce, education could be targeted at those who had the greatest capacity to learn. Those children understood to be intelligent could be channelled towards the academic education that was deemed most appropriate for them. It was these specially gifted children – the 'talented tenth' – who would be able to use their education to have the greatest positive impact on their societies. Everyone else could be directed to vocational or technical education better suited to their abilities.

In the past, decisions about how children should be educated were almost entirely determined by how rich their parents were. Elite schools would admit the worst-performing children as long as their parents could afford to pay, while working-class children with huge intellectual potential were cast out from the education system at the earliest opportunity. But now a new breed of educational psychologists were on hand, promising to provide the tests and technologies to make these judgements in a fair, objective and scientific way.

Just as with vocational testing, the country that most enthusiastically integrated intelligence into the school system was the United States. The American population grew rapidly at the start of the 20th century, with massive immigration from very poor communities in Europe, Latin America and elsewhere. This placed huge new pressures on state governments and school administrators, who had to work out how to integrate a massive and hugely diverse influx of children into a constrained school system. For them, intelligence science and mental testing appeared to be a godsend. Tests could be used to allocate particular children to particular schools, to separate children into different 'streams', and later on to decide who should be allowed to go to college. These testing processes quickly became embedded in the American school system. It was this period that witnessed the birth of the Scholastic Aptitude Test (SAT) and the Educational Testing Service (ETS), which would play such a big role in the American college admission system across the 20th century.[30]

This was not, though, something limited to the United States. The entanglement of education with intelligence and testing in the early 20th century was a global phenomenon. We can get a better sense of the global scale of these changes by looking at Britain and its empire.

In Britain itself, intelligence rose to prominence as the question of educational selection became more important. Selection and assessment hadn't traditionally played much of a role in British education. Education was available to those who could afford it, and educational performance mattered little. Exams weren't introduced at the universities of Oxford and Cambridge until the end of the 18th century.

But the 19th century witnessed a growing interest in meritocracy – the idea that roles in society should be filled by those with the most appropriate aptitudes and qualifications.[31] The idea that someone was qualified for a job merely on the basis of their birth was seen as increasingly out of step with the demands of the modern world. Competitive exams were introduced for recruitment into the Indian Civil Service in 1854, followed by the home Civil Service in 1870. These examinations certainly didn't amount to anything like an intelligence test. They were academic exams, testing a candidate's understanding of subjects like Classics and Mathematics. As a result, they obviously excluded the majority of the population who still didn't have access to secondary education. In this sense they were very similar to the venerable Chinese examination system, which partly inspired them. But they did at least represent an attempt to judge a candidate's power of mind, their talents and their diligence.[32]

From around the 1860s there was also increasing talk in Britain about the importance of an 'educational ladder'. Reformers argued that intelligent boys (not girls, obviously) from poor backgrounds should be identified and given access to advanced education through a system of examinations and scholarships. Liberals and some early socialists thought such a system would be transformative, unleashing the talents of the working classes which had previously been suppressed. More conservative critics were sceptical about the extent of these hidden working-class talents. They preferred the metaphor of an educational 'sieve' which would sift out the unintelligent masses while identifying the few children of genuine ability. But the supporters of both the 'ladder' and the 'sieve' theory of educational selection agreed that the current system wasn't working. Where formal selection did exist it varied widely across the country, mainly involving exams to test knowledge of subjects like Latin rather than a systematic attempt to identify the gifted.

This interest in identifying and supporting the gifted was rooted in some of the big political debates of the era. As with the arguments about mass society and the talented tenth, some believed that the new era of British democracy with its expanded franchise would only work with, in the words of one educational reformer, the 'guidance of brains'.[33] Others saw the education of the intelligent as a solution to the problem of national decline, a particular fear following British struggles during the Boer War at the turn of the century. These ideas were particularly popular among the reform-minded socialist intellectuals in the Fabian Society. For leading Fabians like Beatrice Webb, identifying and promoting the gifted was part of the quest to replace aristocratic amateurism with a society governed by professional experts.[34]

All of these ideas and debates came to a head after the First World War. For the first time, it came to be accepted that most children should progress from elementary education to some kind of secondary schooling after the age of 11. The question then became one of selection. A way needed to be found to select the gifted minority of students who could progress to academic secondary or grammar schools, and potentially then on to university. These represented the 'brains' that would help guide Britain forward. The rest could then be funnelled into some kind of technical or vocational education to prepare them for jobs in industry.

At the same time, it became fashionable to talk about 'general intelligence' as the quality on which these decisions about selection should be made. The phrase echoed Charles Spearman's concept of g, and was undoubtedly influenced by the new work of psychologists on the science of intelligence. But it came to be used among the public in a much wider and less specific sense. Industrialists like George Cadbury (of chocolate fame) lauded general intelligence as the key to national efficiency. The Board of Education tried to work out how to integrate it into the education system. Even adverts for sewing machines jumped on the bandwagon, claiming that the best way to increase general intelligence was through needlework lessons in schools.[35]

Once the existence of 'general intelligence' had become widely accepted, intelligence tests became the obvious way to make decisions about selection. Traditional examinations tested what children had learned. But that wasn't the same as testing for someone's intellectual power and potential. A bright child from a poor background might not enjoy the educational advantages needed to pass a traditional exam. Intelligence tests promised to overcome this problem.

The use of these tests was promoted by a new generation of educational psychologists, led by a man named Cyril Burt. Burt had been appointed as what was probably the world's first official educational psychologist at the London County Council in 1913. Burt was a towering figure in British psychology, and the man who introduced intelligence testing to the country. He first published a British version of the Binet-Simon scale and other tests after the war, and from 1919 they began to be used for scholarship examinations in towns and cities around Britain. Burt helped to popularise the idea of general intelligence and intelligence testing, sitting on numerous government committees and broadcasting regularly on the BBC. Like other psychologists he saw himself as a progressive, bringing objectivity to the education system and opening up opportunities to intelligent children regardless of their background. By the 1930s, intelligence tests were being used in almost every region of the country to make decisions about selection, scholarships and free school places.

But their triumph was only ever partial. Unlike in the United States, where intelligence tests on their own were often accepted as a valid basis for selection decisions, in Britain they were almost always used alongside traditional exams. To win a scholarship or get a place in grammar school, a child would not only need to demonstrate that they had a high IQ, but would also need to show that they could translate some Ovid or solve some algebra. Progress depended on *what* you knew, as much as it did on your raw ability or potential. This reflected an ongoing British ambiguity towards the idea of meritocracy and social mobility. Despite the efforts of reformers like Burt, there were plenty of people in the British educational establishment who were willing to defend class privilege in schools. Their influence helped ensure that that the British system retained (and indeed still retains) many of the elitist traditions of the 19th century.

At the same time as all these changes were happening at home, Britain's educational reach stretched a long way beyond its own borders. The interwar period represented the peak of British imperial power – after it had incorporated new territories following the First World War but before the wave of decolonisation which would be sparked by the Second. Despite the 'civilising mission' rhetoric that had surrounded the imperial project, British governments had shown next to no interest in colonial education prior to the 20th century. Now though, the increasing emphasis on education at home and the pressure from anti-colonial movements abroad forced the issue up the agenda. At a time when education at home was being reshaped by ideas of intelligence and merit, the impact of these same ideas was being felt on education across the empire.

The nature of that impact, however, was very uneven. The closest comparison to the British case was in India, where British interest in the education system had deeper roots than in its colonies elsewhere in the world. It was a place where meritocratic ideas had a long pedigree. Competitive exams for the Indian Civil Service, after all, had been introduced before they were used in Britain itself. It was also a place where Britain had a long-standing interest in cultivating an educated native elite. Given the scale of the country and the limited size of the colonial administration, Britain had always had to rely on an Indian mandarin elite for day-to-day governance.

As a result, educators in India were as interested in the new technologies of intelligence as were their counterparts in Britain.[36] The first people to import such technologies were American missionaries after the First World War, who were inspired by Lewis Terman's research and the US experiences with the army tests to use intelligence testing in their Indian schools. This prompted the British educational commissioner, James Alexander Richey, to launch a new initiative in 1921 to replicate the US Army testing system across Indian schools. Like in the United

States, India was witnessing a huge upsurge in demand for education, particularly secondary education, and testing seemed to offer a way to help deal with the challenge. Richey corresponded with Terman, and tasked Indian researchers with adapting the Stanford-Binet tests for Indian schoolchildren.[37] During the 1920s and 1930s, various Urdu and Hindi translations of intelligence tests appeared, alongside new Indian tests standardised against the scores of local children. These were adopted in schools across the country to assess scholarship applications and make selection decisions.

This represented good propaganda for the colonial administration. British officials claimed that the use of intelligence testing demonstrated their commitment to merit and objectivity, replacing cronyism and corruption with the principles of open competition. It also helped to showcase the British fight against the caste system, with British rulers arguing they were replacing the old hierarchy of birth with a new hierarchy of intellect.

But talent wasn't everything as far as the British were concerned. In line with the traditions of missionary education which had emphasised the training of character, they were as much interested in moral education as they were education of the intellect. The new Indian elites, from a British perspective, needed an education advanced enough to equip them for the demands of modern government. But they also needed an education that would inculcate respect for British values, and would thus shield them from the lure of anti-colonial politics. British officials tended to see mental and moral development as two sides of the same coin. By selecting children with the greatest mental capacity, therefore, they believed they would also be identifying those with the moral capacity needed to loyally serve the enlightened interests of British rule. The flip side of this idea was the assumption that only the irredeemably unintelligent would embrace the cause of anti-colonialism.

If intelligence and testing became increasingly important in the Indian education system, its influence in Britain's African colonies was much less evident. This was partly because of the more overtly racist ideas about African intelligence held by British settlers and administrators. Kenya, for example, saw the development of a particularly virulent eugenics movement during the 1920s and 1930s, led by White settlers and tied to similar movements in Australia and South Africa. White eugenicists argued that low intelligence made Africans effectively 'uneducatable', and developed local versions of intelligence tests to try to prove it. Their ideas were widely shared among the settler community, and were amplified in Britain by the Eugenics Society.[38]

Most British psychologists and educationalists were sceptical about these overtly racist claims.[39] In Britain, such arguments were attacked by intelligence researchers like Cyril Burt. More liberal colonial administrators in Kenya argued that intelligence testing should be used to select talented Africans for advanced education, rather than attempting to prove African intellectual inferiority. But even these critics firmly believed that the average intellectual level of Africans was below that of Europeans, and that the racial division between those doing 'mental' and 'manual' jobs needed to be maintained.

This reflected a wider set of beliefs about African education among British colonial rulers. Colonial governments had shown almost no interest in education prior to the First World War, and the Western-style schools that provided basic education for a tiny minority of Africans were run entirely by missionaries. In the 1920s, it was these missionary groups that first started to develop systematic policies towards education in Africa, and which in turn prompted the British government to adopt its own plans. The most striking thing about this process was the inspiration it took from debates about Black education in the United States.

The educational link between colonial Africa and the United States was underpinned by a study conducted between 1920 and 1922 by the Phelps-Stokes Fund, an American philanthropic organisation whose work focused on the education of African Americans. The study was commissioned by European and American missionary groups with the co-operation of colonial governments, and was led by a White American sociologist named Thomas Jesse Jones. Jones's explicit aim was to apply to African education the same principles that he and the fund had been advocating for African-American education in the US. To understand where these ideas came from, we have to return to W.E.B. Du Bois and his arguments about the talented tenth.

Du Bois had aimed his arguments against the approach of Booker T. Washington. Washington's famous Tuskegee Institute, founded in Alabama in 1881, promoted industrial education as the best approach to Black educational advancement. Thomas Jesse Jones and the Phelps-Stokes Fund were big supporters of Washington and Tuskegee, arguing that this kind of education prepared African Americans to live an 'upright, useful, and economically productive life'.[40] Du Bois had seen this as a recipe for maintaining Black subjugation, and argued that true freedom and progress could only be achieved through the advanced liberal education of a Black elite. But Jones and the Phelps Foundation argued that the vocational system they thought so appropriate for Black Americans was also that best suited to the African context. For them, education in Africa needed to be 'adapted' to the specific needs of African colonial society.

British colonial rulers viewed African education as 'a not wholly similar but an analogous problem', as one official put it, to the education of Black Americans in the US South.[41] They accepted Jones's argument that the Tuskegee vocational model was ideally suited to the African context. As the official 1925 British policy on education in Africa put it, education should be adapted to 'the mentality, aptitudes, occupations and traditions' of African societies.[42] What that meant in practice was that education should be focused on the technical training that would be needed to form productive workers, rather than on the traditional academic subjects that were deemed appropriate for most children in Britain. British officials argued that African schools should focus on agricultural skills, which they saw as the key to the development of African economies. And they were particularly keen to ensure that Western-style education wouldn't alienate children from their rural African communities, luring them towards the cities or making them discontented with their social and economic circumstances.

But, like in India, British officials also believed that some children needed to be given access to advanced education. It was this elite minority of African children who would go on to staff the local colonial administrations and become the teachers needed to educate the next generation. To produce these educated African elites, they promoted the creation of colleges, later to become universities, in places like Ibadan in Nigeria and Makerere in Uganda.

This definitely didn't mean that British officials were interested in the uplift of an African 'talented tenth' along the lines that Du Bois was advocating for Black Americans. One of the reasons, they believed, that such a strong anti-colonial movement had emerged in India was that British rulers had been too lax in allowing Indians access to higher education, creating a discontented intellectual elite who became drawn to nationalism.[43] They were determined not to make the same mistake in Africa. Their vision of elite education, therefore, focused as much on moral as on intellectual training. They wanted to create 'trustworthy' and 'public-spirited' elites who could lead their people, but who wouldn't rock the colonial boat.[44]

These policies attracted many of the same criticisms that Du Bois had been aiming at the Tuskegee vocational educational model in the United States. Parmenas Mockerie, a Kenyan teacher who had been trained at Makerere College and was an associate of the anti-colonial leader Jomo Kenyatta, argued in 1934 that 'Africans ... are conscious of the necessity of having a sound education similar to that given to children of other races.' Europeans, he complained, said that Africans couldn't contribute intelligently to democratic government because they weren't sufficiently educated. But they then turned around and claimed that Africans couldn't be given the advanced education needed to be good democratic citizens because it wasn't compatible with their 'mental capacity'.[45] Africans were thus stuck in a colonial catch-22: too unintelligent to be educated, and not educated enough to be free.

These criticisms fell on deaf ears, and the British approach to African education remained largely unchanged right up to the final years of colonial rule in the 1950s. Education remained 'adapted', focusing on agriculture and technical skills, and designed to produce productive workers contented with their social circumstances. So uninterested were the British in advanced education that, by the early 1950s, only around 0.1 per cent of children in Britain's East African colonies were finishing secondary school.[46]

Taken to its extreme, mental engineering didn't mean just changing human society, but changing human *biology*. As we've already seen, the idea of eugenics was pioneered by

Francis Galton, the man who kickstarted the scientific study of intelligence. In the early 20th century, intelligence became central to the eugenicist project of 'improving' the human race.

Many of the leading intelligence scientists of the era were ardent eugenicists. They tended to make two distinct arguments about intelligence and human betterment. The first was that the overall intellectual quality of the human race could be improved by encouraging the most intelligent to have more children. We'll hear more about this idea later in the book. The second was that the same improvement could be achieved by limiting the breeding of the least intelligent. This would liberate society from the burden of caring for and managing the behaviour of the people who were, eugenicists believed, the main cause of crime, delinquency and squalor. Lewis Terman, one of the most active eugenicist psychologists, claimed that 'all feeble-minded are at least potential criminals'.[47] His colleague Edward Thorndike argued that the 'one sure service ... which the inferior and vicious can perform is to prevent their genes from survival'.[48]

The simplest way to achieve this was through sterilisation of those deemed to be feeble-minded or mentally unfit. US states began to introduce forced sterilisation laws from 1907, with Terman's California the most enthusiastic sterilising state. By the 1930s, over half of all US states had passed compulsory sterilisation laws, many of which remained on the books until the 1970s. Added to this were new policies to prevent the immigration of those with low intelligence or mental disorders, aided by the use of intelligence testing in immigration processes. These sterilisation policies and other eugenic measures spread around the world in the 1920s and 1930s. Eugenicists even began to think about the need to manage the size and quality of the global population, often driven by fears that White Western civilisation was going to be drowned in a rising tide of non-White races.[49]

These policies reached their logical extreme in Nazi Germany. Unlike in the United States, Nazi eugenics had little interest in the notion of high intelligence. The psychological study of intelligence and giftedness had thrived in Germany during the first decades of the 20th century, but declined with the arrival of the Nazi regime. The Nazi understanding of racial hierarchy certainly shared the belief prevalent in other Western societies that non-White races lacked intelligence. On the other hand, Nazi anti-Semitism also encompassed the idea of the Jews as a uniquely clever race, even if it held them incapable of true genius or intellectual greatness.[50] Indeed, the idea of Jewish cunning was used to justify their supposed threat.

The German race was also celebrated for its talents and creativity, and some Nazi leaders spoke about the importance of the mental powers of leading minds in preserving racial strength. But the Nazi understanding of Aryan racial superiority was never limited to the notion of high intelligence. In elite Nazi boarding schools, for example, potential recruits were tested for their intelligence and academic ability, and were expected to perform well. But their mental abilities were less important than their physical vigour, their courage and their racial purity.[51] It was these qualities that were most important for shaping future Nazi leaders. 'When it comes to the filling of leading posts in the state and party,' Hitler argued in 1939, 'character should be valued more highly than so-called academic or supposed intellectual qualifications.'[52]

Where intelligence was important to Nazi policy was in the treatment of those whose low intelligence was judged to pose a threat to the health of the race, or to prevent them from being productive members of society. Germany had long had an active eugenics movement, and previous governments had adopted eugenic measures. But the Nazi understanding of racial hygiene represented a step change. Inspired in part by sterilisation laws in the US, the new Nazi regime passed the 1933 *Law for the Prevention of Offspring with Hereditary Diseases*, mandating the mass sterilisation of people with disabilities and those from other 'asocial' groups. This was followed by the wartime T4 euthanasia programme which oversaw the systematic murder of a quarter of a million adults and children institutionalised with physical or mental disabilities.

There were plenty of people around the world at the start of the 20th century who challenged the new understanding of intelligence that was being used to reshape society, just like there were plenty of people who challenged eugenics and other forms of scientific racism. In the US, for example, the well-known journalist Walter Lippmann launched a scathing critique of Lewis Terman and the intelligence testers in a series of articles in 1923. He denounced the tests as little more than a series of 'stunts' and 'puzzles' which psychologists had put together without any real idea of what intelligence was.[53] All they really did, he claimed, was to classify people, to arrange a group into a hierarchy from best to worst according to their ability to complete these arbitrary sets of puzzles.

But let's remember the story of Margaret, who we began this chapter with. She and her friends may well have agreed with the kind of criticisms of the tests being advanced by Walter Lippmann. Unlike Lippmann, though, they weren't in a position to air these criticisms in the national press. There wasn't even any way for Margaret to challenge the negative result of the test she'd taken. All she was left with was the label – the official decision that she was too 'unintelligent' to join her sister in the army. Supporters of meritocracy argued that society's respect should be transferred from the well-born to the people of ability. They had little concern about the respect due to those who fell at the hurdles thrown up by their brave new meritocratic society.

People like Margaret weren't the social engineers, the experts who decided that intelligence and intelligence science should be used to better manage society. They were the ones *being engineered*. They weren't among those who were judged to be highly intelligent and who benefitted from these new ideas and technologies, of which there quite a few. Thankfully, they weren't among the hundreds of thousands of people around the world killed or mutilated on the grounds of their supposed low intelligence as part of state eugenic programmes. But their experience was perhaps the most representative of the way ordinary people encountered the new vision of intelligence that emerged in the early 20th century. They were the people who were discarded or rejected on the basis of judgements about their intelligence, who found their dreams dashed or their life choices limited. When we think about the history of intelligence and mental engineering, these are the experiences we should try to remember.

5

Mensa

If all the good people were clever,
And all clever people were good,
The world would be nicer than ever
We thought that it possibly could.

But somehow 'tis seldom or never
The two hit it off as they should.
The good are so harsh to the clever
The clever so rude to the good.

Elizabeth Wordsworth

IN A 1999 EPISODE OF *The Simpsons*, Lisa Simpson is invited to join the Springfield chapter of Mensa, the high IQ society. The group engage in highbrow discussions and lament the anti-intellectualism of their fellow citizens. But after Mayor Quimby skips town, they find themselves in charge. 'With our superior intellects,' Lisa declares, 'we can rebuild this city on a foundation of reason and enlightenment. We can turn Springfield into a utopia!'

The intellectual junta set about reforming the city along rational lines, redesigning the traffic system and introducing metric time. But the group soon descends into infighting. Its proposals to ban all sports, limit breeding and promote broccoli juice draw the ire of the townsfolk. Stephen Hawking, in the first of many *Simpsons* guest appearances, arrives to express his disappointment, and a riot breaks out ('Come on you idiots, we're taking back this town!', Homer declares). Hawking flies Lisa to safety on his wheelchair helicopter, and life returns to normal.

The episode gently mocked Mensa and the kind of people who joined it, but it also demonstrated how much cultural resonance the organisation had come to enjoy. By the end of the century it had become an almost universally recognised cultural phenomenon, a popular shorthand for the idea of high intelligence. Stephen Hawking may never have signed up, but lots of serious figures did – from the renowned American architect and intellectual Buckminster Fuller to the science fiction pioneer Isaac Asimov.

Despite this high-profile and cultural footprint, Mensa has had very little practical impact on society. It hasn't influenced government policy, shaped public opinion or successfully conspired to further its own aims. In fact, it has often been regarded by the public as something between a harmless society of misfits and a slightly distasteful club for the smug and self-regarding.[1]

But it does reveal something important about the way intelligence came to be understood from the middle of the 20th century. The idea of forming a society solely for people with high IQs was remarkable enough in itself, especially given how new the concept of IQ was. But the fact that such an organisation could grow from its origins as a club of misfits and eccentrics in post-war England into a mass organisation with tens of thousands of members in every corner of the globe shows that the new ideas about intelligence that had emerged and spread at the start of the 20th century had gained real traction.

Mensa also reveals some of the complexities and contradictions which came to surround the idea of intelligence in the second half of the 20th century. Mensa members hoped that they would be able to mobilise their intelligence in the service of humanity, to make the world 'nicer than ever', as Elizabeth Wordsworth's poem put it. This belief reflected the increasingly popular idea of intelligence as a resource which societies could use to drive progress and development.

But as the poem also suggests, these hopes weren't always easy to realise. Mensa members had a complicated relationship with the non-Mensa population, sometimes insecure and self-conscious about their differences, sometimes rude and dismissive towards those they deemed less intelligent. Mensa members campaigned to improve the educational chances of underprivileged children, but they also dabbled in eugenics and embraced neoliberalism as a solution to what they saw as the problem of excessive equality. The flip side of thinking that intelligence could drive progress was the idea that society, just like in the *Simpsons* episode, should be led by a highly intelligent elite.

Mensa: the early years

Roland Berrill was, to put it mildly, eccentric. A larger-than-life figure with a goatee and a twirling handlebar moustache, Berrill dedicated much of his life to promoting various esoteric beliefs. He had been born in Australia at the end of the 19th century, the second son of a rich British family. He was educated at public school in Britain and later went to Oxford before sitting for the Bar in London, although his family wealth meant that he never really had to work. His passions included astrology, phrenology and Dianetics (L. Ron Hubbard's pseudoscientific version of psychoanalysis that was a forerunner of Scientology). And he was active in the male dress reform movement, which called for men to be able to wear loose and colourful clothing like women could.

But his most important achievement, and the reason he's of interest to us, was the founding of Mensa in 1946.

There are different accounts of where the idea for a high IQ association first came from. Some point to Lancelot Ware, a British barrister whom Berrill met on a train in 1945 and who first sparked his interest in IQ testing. Others say it was the leading British intelligence scientist and educational psychologist Cyril Burt, who first discussed the idea in a BBC radio series the same year. But although both men went on to hold senior roles within Mensa, the organisation was very much Berrill's creation, and was led and almost entirely dominated by him for the first five years of its existence.

Roland Berrill. Reproduced by kind permission of British Mensa

Those five years, it has to be said, weren't particularly successful. Membership was limited to a few dozen people, almost all of whom were acquaintances of Berrill or Ware. They would meet every few months in hotel dining rooms and restaurants around London, and shared ideas through a monthly journal. The fact that Berrill founded and funded the organisation, as well as editing its journal, meant that he had almost unlimited control over the way it functioned. This inevitably meant that early Mensa reflected his personality. While many of the initial members found him charismatic and engaging, others were put off by the various eccentricities he foisted upon them.

These included his habit of using the Mensa journal to harangue members for their laziness, and his insistence that Mensa gatherings should be presided over by a female member who functioned as a kind of ceremonial Mensa queen (the '*Corps d'Esprit*'), and who, often with great reluctance, was forced to sit through meetings on a throne dressed in special robes while being toasted by other members. Berrill also allegedly insisted on being present during the undressing and robing of the *Corps d'Esprit*, which casts a more sinister light on some of his 'eccentricities'.[2]

None of this was a model for a thriving or sustainable organisation, and by 1952 Berrill had been ousted. At that stage Mensa had slightly more than 300 members, and seemed, in the view of many of those members, destined to die a slow and largely unmourned death. But, against all the odds, its fortunes were revived. Under the leadership of another charismatic and luxuriously bearded member, Victor Serebriakoff, Mensa began to thrive.

By the end of the 1950s it had grown to over 1,000 members. During the next decade it expanded beyond Britain, recruiting 15,000 members in over 15 countries, with the majority of those members in the United States. By the end of the 1980s that number had reached over 70,000. Today, Mensa boasts over 140,000 members in more than 100 countries around the world. It's an organisation almost everyone has heard of, and which is in many ways synonymous with the idea of high intelligence.

The number of members alone suggests that it's an organisation whose history has to be taken seriously.

But it's also worth looking at what lies behind those numbers. Mensa membership has always had quite a high turnover, as new members join and old members (perhaps finding that being a Mensan isn't all it's cracked up to be) drop out. That means that over the years there have been far more members than the current global total of 140,000. And Mensa only admits those who score in the top 2 per cent of the population in IQ tests. There are no published figures for how many people have taken IQ tests with Mensa, but various estimates over the years have suggested that only 1 in 10 people who do so reach the required score. That means we can be fairly confident that, since the end of the Second World War, *millions* of people around the world have chosen to take an IQ test through Mensa, and that hundreds of thousands of those have gone on to join the organisation, however briefly.

IQ tests, as we have seen, had been used increasingly widely in many countries since the First World War. But their use had almost always been confined to institutional settings – schools, the armed forces, research institutions or medical facilities. They had been given *to* people, for particular purposes. Now though, people were *volunteering* to take these tests (and indeed paying for the privilege), for no other reason than to find out if they scored highly, and if they did to give them the chance of joining a society for people with whom they had nothing in common except for their high IQ. Mensa's success heralded a new era in the popular embrace of intelligence.

Being a Mensan

So why did so many people decide they wanted to join Mensa? It's not necessarily an easy question to answer. We can't assume that everyone who either took a Mensa IQ test or decided to join the organisation was motivated by the same impulses. And Mensans have sometimes been shy or embarrassed about their membership, reluctant to discuss it openly with outsiders. We can, however, discern some common reasons for deciding to join.

The main pull for most members was social. From its very beginning, a key part of Mensa's work was providing opportunities for its members to socialise together, organising regular dinners, drinks receptions and residential weekends. In later years these were complemented by annual meetings, regional seminars, week-long residential conferences and other activities. The first Mensa *Booklet of Common Interests and Unique Skills* was published in 1948, listing the contact details, qualifications and interests of all members to encourage connections between like-minded people.[3] These later evolved into Special Interest Groups (SIGs), which members could set up around any conceivable shared activity or topic, and which remain at the heart of Mensa activities today. The entrepreneur, inventor and long-time Chair of British Mensa, Clive Sinclair, reflected that most of his close friendships had been made through Mensa, and many other members seem to have had a similar experience.[4]

Mensa, of course, isn't the only club that encourages socialising, and its members could presumably have made friends or gone to dinners and conferences just as easily by joining societies for knitting, kayaking or pigeon fancying. There must, therefore, have been some particular appeal to the idea of socialising exclusively with people who scored highly on an IQ test.

Victor Serebriakoff, the man who succeeded Berrill and oversaw Mensa's massive global growth, argued in the Mensa journal in 1957 that members were drawn to the opportunity for 'serious discussion' with people who were 'interested in everything'.[5] From an early stage, member meetings were organised around talks by leading academics and discussion of 'serious' topics, particularly computing and artificial intelligence, which were of keen interest to many members from the 1960s. For those who hadn't had the chance to attend university, Mensa often offered the only opportunity for the kind of intellectual discussion they craved.

That's not to say that discussions in Mensa were always so rational or highbrow. Shortly after joining in 1960, a new member from South London wrote to the Mensa journal to complain that when he joined Mensa he assumed it would be 'free from mention of such things as extra-sensory perception, witches, and telepathy. I have been astonished to find these things to be a frequent topic of conversation.'[6] Indeed, early surveys suggested that a large minority of Mensa members were happy to declare their belief in faith healing, Scientology and the occult.[7] But a lot of people seemed to have quite enjoyed this mix of the serious and the slightly odd. As another early member argued, the eccentrics in Mensa were 'enough only to spice the mixture, not to overflavour it'.[8]

For some, Mensa also offered an opportunity to overcome the self-consciousness and social difficulties they faced in the real world. The social rejection apparently faced by intelligent people was a consistent topic of discussion in Mensa meetings and publications. Sometimes this was put down to their predilection for odd or eccentric beliefs. But many members argued that the intelligent naturally tended towards introversion, exacerbated by a lifetime of feeling the need to hide their intelligence from friends and family at the risk of seeming too clever.[9] Journalists who reported on Mensa often noted people's reluctance to reveal their membership publicly and risk being seen as pretentious. Taking an IQ test and scoring highly enough to join Mensa was, for many, a way of boosting their self-confidence, even if it was one they felt slightly embarrassed by.[10]

But this self-consciousness could easily blur into a sense of superiority. The very first Mensa magazine published in 1947 argued that all people can be very quickly classified into two categories, 'the intelligent ... and the rest', and that 'an alarmingly small number can be firmly placed in category one!'[11] For Victor Serebriakoff, one of the benefits of a Mensa meeting was that there was 'no need for the tentative preliminaries which the intelligent person must normally go through in order to assess the mental level of a new acquaintance: the newcomer can embark on discussion of any subject with the assurance that his hearer will understand the points he makes and know when they are proved'.[12] If members struggled to find engaging, intelligent conversation in normal life, this attitude suggested, it was because their intelligence placed them above the understanding of ordinary people.

Although Mensa's membership has included many successful businesspeople, artists and academics, there has been a persistent idea both inside and outside the organisation that it has attracted a disproportionate number of misfits and underachievers, for whom a high IQ score offered the validation they lacked from the outside world. It's also worth noting that Mensa membership has consistently remained around two-thirds male, a fact attributed by some to men's greater desire for external validation. As one of the earliest newspaper articles on Mensa commented archly in 1961, Mensa was 'an organisation for those whose I.Q. puts them among the top 2 per cent of Britain – or those who feel the need for proving it'.[13]

Intelligence in the service of humanity

Mensa always aspired to be more than a social club for the bright and underappreciated. It aimed, as was later written into its constitution, to foster intelligence 'for the benefit of humanity'. While this sounds like a lovely idea in theory, what it meant in practice was trickier to pin down.

Roland Berrill saw Mensa as a kind of brains trust. His vision of the organisation was explicitly aristocratic – he thought it should be confined to a few hundred people, and was happy for most of them to come from the friends, family and acquaintances of the original members.[14] This aristocracy of intelligence, he argued, should serve to 'clean the window through which Power looks out to the world'. With that in mind, he dedicated much of his effort in the early years of the organisation to running what he called 'interrogatories'. These were postal opinion surveys of members, the results of which he published in each quarterly magazine. The questions ranged from the philosophical (*Do you believe in life after death?*) to the political (*Should Britain remain in Berlin?*). And Berrill clearly hoped that the results would be of practical use to the government of the day, chastising members that he wanted answers of sufficient merit to be sent to the prime minister, and going so far as sending the results of his survey on crime and punishment to the Home Office.[15]

Alas, Berrill's unsolicited policy advice appears to have fallen on stony ground. 'There is not,' as Victor Serebriakoff lamented in 1961, 'unfortunately, a queue of University departments waiting to avail themselves of our services, nor of politicians or civil servants waiting to translate our findings into legislation or administrative decisions.'[16] Serebriakoff blamed this on the lack of agreement among Mensa members. In the early years of the organisation it had been hoped that their shared intelligence would allow members to reach a uniquely wise and well-considered consensus on the burning issues of the day. But Berrill's interrogatories only succeeded in demonstrating that Mensans held as wide a range of opinions on any given subject as any other random cross-section of the population, if not wider.

Ever since this early realisation there have been periodic attempts to revise the idea of Mensa as a resource for governance, from various short-lived Mensa think tanks to the occasional meeting with politicians to discuss pet topics such as gifted education. But with a few minor exceptions, mainly outside of the US and UK, Mensa has signally failed to exert any meaningful influence on government policy.

It did have slightly more success with universities. If the lack of consensus among Mensans undermined their usefulness as a resource for governments, then at least psychologists, sociologists or other academics might find use for an accessible group of high IQ holders as part of their research. By the late 1950s Mensa had forged strong links with a number of researchers, particularly in London's centres for psychological research at University College London and the Maudsley Institute of Psychiatry. Mensa members participated in studies into intelligence and education, the measurement of personality and the validation of IQ tests. Hans Eysenck, one of the world's leading psychologists who would later go on to be a high-profile science communicator and controversial advocate for ideas about IQ, race and hereditary, was one of the closest early collaborators, undertaking a study of Mensa personalities in 1958 and speaking regularly at Mensa events.[17]

The most concrete way in which Mensa sought to promote intelligence in the service of humanity was the field of gifted education. This was another concern that was evident right from the earliest years of Mensa. The first Mensa magazine, in 1947, for example, reported on concern in educational circles about the lack of adequate teaching for 'exceptional' children.[18] Members felt that – as former gifted children themselves – they were ideally placed to understand the problems these children faced and to advocate for better provision.

By the late 1950s Mensa had organised research into the difficulties facing intelligent children, and began to discuss the idea of special schools for those with high IQs (potentially staffed by an all-Mensa teaching body). British members were in contact with gifted education pioneers in the United States, where the movement was strongest, as well as in countries around the world, stretching from Italy to India. Victor Serebriakoff claimed to have played a role in the foundation of Britain's National Association for Gifted Children by providing materials from their US counterparts in the 1960s.[19]

Mensa's involvement with the gifted children's movement really took off on both sides of the Atlantic from the 1970s, when it developed official posts for experts in gifted children's education, and began offering scholarships and mentoring to children. In 1989, British Mensa's new Foundation for Gifted Children launched a national advertising drive to find the 1,000 cleverest pupils in Britain, printing mini IQ quizzes in newspapers and asking the parents of children who scored well to apply for a full IQ test.[20] The response was huge, and Victor Serebriakoff was invited to meet with the Department for Education to discuss potential policy changes (one of those rare examples of Mensa engagement with policy-making). In 1990, British Mensa went as far as setting up a new private school in Southwest London aimed at high-IQ pupils.[21]

Gifted children's education remains a major part of Mensa's work today, and particularly of its publicity machine. Although it didn't admit children as members until the 1980s, the vast majority of the press coverage it now receives relates to children, sometimes as young as three or four, who have scored highly enough on IQ tests to become members.

Underpinning all of these projects was the idea that the highly intelligent *could* genuinely benefit humanity, that the successful cultivation of intelligence had the potential to make a difference to modern societies. This was one of Mensa's core beliefs, and arguably the one that had the widest public traction.

For Cyril Burt, the famous British psychologist and first President of Mensa, the importance of intelligence was rooted in the unprecedented complexity of the modern world. 'Now almost all the important questions of modern life,' he wrote in 1966, 'have become so intricate and tangled that only persons of first-class intelligence can unravel them and reduce them to manageable proportions.'[22] In the age of atomic power and space travel, of globalisation and geopolitical upheaval, all societies needed the highly intelligent to solve the puzzle of how to drive development and progress. By bringing together those of high intelligence,

Burt argued, Mensa could make a vital contribution 'to the advancement both of civilized and under-developed countries, and (it is surely not too much to say), to the peace and progress of mankind'.[23]

But this faith in the power of intelligence coexisted with another set of beliefs about the importance of inequality and the need for a highly intelligent elite. Throughout the history of Mensa's gifted children work it was clear that, for many members at least, supporting gifted education represented a way to fight what they saw as the misguided egalitarianism of the modern world.

This was particularly the case for Victor Serebriakoff. At the end of the Second World War, Britain had introduced a new secondary education system in which pupils were to be divided at the age of 11 according to the results of the so-called '11+' exam. Those who performed best in these exams attended more academically focused grammar schools, and the rest were sent to technical or secondary modern schools, where they would receive a more 'practical' education. This system started to change in the 1950s when local councils began scrapping the 11+ and combining these different types of institutions into comprehensive schools, a process which was later promoted across all of England and Wales by the Labour government after 1965. It was a change driven partly by pressure from parents who wanted good community schools for all children, and partly by educational experts who increasingly criticised the outcomes of the 11+ and the injustices of the selective system.[24] But it was also deeply controversial.

Serebriakoff was a fierce critic of comprehensive schools. He argued that the selective system of the 1940s and 1950s had succeeded in improving access to high-quality education for intelligent children from working-class backgrounds. The turn against selection in the 1960s, he claimed, was the result of an unholy alliance between egalitarian zealots on the one hand and middle-class parents on the other, who had been alarmed by the new competition from these intelligent working-class children and had therefore turned against selection of any kind.[25] Together they had succeeded in undermining Britain's world-leading secondary education system and denying Britain the intellectual talent it needed to develop in the modern era of advanced technology and the information revolution. Teachers had been taught to despise all differences in educational achievement and had therefore abandoned the most able children in order to concentrate on the least able. 'They stop pursuing excellence,' he lamented, 'and go for equal mediocrity.'[26]

Despite his apparent concern for the opportunities of working-class children, the educational policies that Serebriakoff and his Mensa colleagues advocated seemed squarely aimed at defending the traditional class-stratified features of the British education system. In 1972, Serebriakoff launched a new Mensa group to lobby the government in favour of education for the gifted. In the Mensa magazine, he urged members to defend top public schools such as Winchester and Eton against the supposed threat from Labour education reformers, arguing that 'Mensa must do its best to propagate the idea of an elite.'[27]

These ideas were even more evident in the United States. One of the American Mensans who was most active in the gifted education movement was Nathaniel Weyl. Weyl (who was also a member of a group called *M/M*, which was restricted to those with the highest 2 per cent of IQ scores *within* Mensa) worked closely with Serbriakoff on gifted education projects. These included the International Foundation for Gifted Children, which he founded in the 1960s; projects to examine levels of giftedness in different countries; fundraising and support for gifted education projects in Italy, Ecuador and Korea; lobbying the Nixon administration on the adoption of an international gifted education policy; and even a proposal for an international adoption agency to match Mensa members to gifted orphans around the world.[28]

Weyl's interest in gifted children was rooted in his politics, which combined fierce anti-communism with an obsession with IQ, race and eugenics. An economist by profession, he published widely on all these topics between the 1960s and the 1980s, including numerous books and articles on racial differences in intelligence, and on the supposed intellectual superiority of Jews and White Rhodesians. He collaborated with notorious race theorists like Arthur Jensen, William Shockley and Richard Lynn, and published regularly in the race science journal, *Mankind Quarterly*.[29]

Weyl's interest in gifted education was rooted in his belief that progress throughout history had always been driven by a highly intelligent elite. Identifying and promoting such an elite, he believed, was the best way of driving forward global development. On the other hand, he argued, underdeveloped countries were underdeveloped because they lacked sufficient numbers of gifted individuals, whether because of genetic or civilisational failings (as he argued was the case with Arab societies in the Middle East and Black societies in southern Africa), or because their intellectual elites had been wiped out by communist or anti-colonial movements envious of the gifted.[30]

Mensa politics

A club offering social space for like-minded misfits who seek to use their intelligence to benefit humanity sounds fairly unobjectionable. But as the example of Nathaniel Weyl suggests, there was a darker side to the politics of Mensa.

This is not something that Mensa has ever been keen to acknowledge. In fact, it has always worked hard to present itself as fundamentally non-political. As early as 1957 it adopted a series of policies declaring that Mensa would not do anything to alienate or exclude anyone, that it would take no political stance or actions, and that, while its members may have their own opinions on whatever topic they liked, 'Mensa has no opinion'.[31] These policies remain in force today.

But they haven't succeeded in protecting Mensa from suspicion and criticism of its politics. Throughout its history, most of the press coverage about Mensa has been either gently mocking or openly disdainful. The obvious criticism of Mensa is that it's inherently elitist, and that its members must, by definition, think they're better than everyone else. But there have also been persistent suspicions that Mensa's bland public image masks something more politically sinister. In 2018, the American comedian Jamie Loftus signed up to Mensa and discovered that it was host to a thriving community of far-right trolls who spent much of their time spewing hate on online Mensa forums, and doing their best to make life as unpleasant as possible for other members who disagreed with them.

So what's the truth? Is Mensa really non-political? And if it's not, what *are* its politics? One obvious defence of Mensa's claims to political neutrality is the fact that its members naturally hold a range of political opinions precisely because they have nothing in common beyond their high IQ. (This isn't entirely true, despite being an argument often made by Mensans – they also share the fact that they've (a) chosen to take an IQ test, and (b) decided to join Mensa.) This apparent political diversity was evident from the very early history of the organisation, when, despite Roland Berrill's aristocratic yearnings, a not insubstantial minority of members were happy to declare their faith in communism.[32]

But an examination of Mensa's history makes it hard to ignore the varieties of right-wing politics which keep bubbling up to the surface. We've already seen some of the anti-egalitarian ideas that underpinned Mensa's work on gifted education. And the case of Nathaniel Weyl points us towards Mensa's persistent history of entanglements with the idea of eugenics.

This interest in eugenics was evident right from the beginning of the organisation. One of Berrill's early interrogatories canvassed members' opinions on the topic. And one of the first external speakers invited to address a Mensa gathering in 1955 was Cedric Carter, a British geneticist who was then the secretary of the British Eugenics Society. Carter warned members that advances in modern medicine were perpetuating the 'heredity of sub-normal health'. His proposals to address this problem included discouraging parents who were 'ill-endowed genetically' from having children, while encouraging larger families among those with a 'good genetic endowment'.[33]

By the standards of the time, these views probably weren't perceived to be as extreme as we'd now assume them to be. Although we tend to think that eugenics was universally discredited by its association with Nazism, it remained a relatively mainstream idea until well into the 1950s. The famous British biologist Julian Huxley, for example, openly advocated for global eugenic policies in his role as the first post-war leader of UNESCO.[34]

But controversies over eugenic ideas seemed to dog Mensa long after they had become publicly taboo. These controversies were often related to the concept of dysgenics – the fear, such as that expressed by Cedric

Carter, that conditions in the modern world were undermining human genetic quality. In Victor Serebriakoff's first book about Mensa, he argued that average intelligence in Britain was declining because higher IQs were more prevalent among the upper classes, who had a lower birthrate than other social strata.[35]

Another figure concerned about dysgenics was Robert Graham, a Californian millionaire who had made his fortune manufacturing contact lenses. Graham believed that the way to prevent the genetic degeneration of the human race was for intelligent people to have more children. His solution to this problem was to create a sperm bank for the highly intelligent, originally envisaged as being limited to Nobel Prize winners, and to use it to breed intelligent babies. His *Repository for Germinal Choice* only managed to extract the sperm of one Nobel winner – the computing pioneer turned notorious race theorist William Shockley, who we will discuss in a later chapter. But it did succeed in building up a collection of sperm from college maths and science students around California.

Graham was not himself a Mensa member, but in 1980 reports emerged claiming that he had approached female Mensa members to find volunteers to be inseminated with his Nobel-grade sperm.[36] In 1983 Graham admitted that he had indeed tried this strategy, albeit with little success. In the same year he was invited to address Mensa's major annual gathering at Cambridge University to explain his ideas.[37] And in 1996, his sperm bank still apparently going strong, he teamed up American Mensa's Eugenics Special Interest Group (which Nathanial Weyl was involved with) to advertise for sperm donors in the Mensa magazine.[38] Eugenic ideas seemed to be particularly popular in Mensa at that time. Just the year before, the editor of the Los Angeles Mensa magazine was fired after publishing an article by two members recommending the 'euthanasia' of the mentally disabled, homeless and elderly.[39]

This does not mean, of course, either that most Mensans were ardent eugenicists, or that eugenics was a core part of the organisation's belief system. But we *can* identify a broader set of political beliefs shared by many leading Mensans from the 1940s onwards, which were built around the idea of free markets, meritocracy and anti-egalitarianism.

When Mensa was founded in the aftermath of the Second World War its members often suggested that the organisation was at odds with the egalitarian spirit of the era. Reporting on the 1958 annual meeting, Mensa's magazine argued that 'the idea of a self-conscious and dedicated elite of people whose ability has been scientifically checked is one which excited instant revulsion in all educated people in this egalitarian age'.[40] Mensa complaints about egalitarianism often verged into more explicit anti-socialism. In the 1960s, Victor Serebriakoff complained that the lamentable disregard for intelligence in modern society could be explained by the fact that the country had been for so long 'dominated politically by an egalitarian ethos with Fabian socialism as its extreme form'.[41]

Mensa's big period of expansion came between the 1970s and the 1990s, driven by a rapidly growing membership in the United States. This period witnessed a broad turn against the social democratic egalitarianism of the post-war era in western Europe and North America. In its place came the embrace of neoliberalism – free market economics and government reforms, combined with a cultural turn towards ideas of economic freedom and entrepreneurship, and an increasing social acceptance of inequality. The link between these political trends and the growth of Mensa isn't straightforward; its growth may have been driven by better marketing techniques as much as by the emergence of new social attitudes. But many in Mensa understood its expansion as a reflection of a renewed popular acceptance of inequality, hierarchy and elites – or, in the words of Victor Serebriakoff, a 'counter-thrust to the over-egalitarian trend'.[42]

Key Mensa figures during this period were notable supporters of neoliberal notions of meritocracy and social hierarchy. This was particularly the case in Britain. During the 1980s the organisation was dominated by two figures: Clive Sinclair, who chaired British Mensa, and Madsen Pirie, its international secretary. Sinclair was a prominent British businessman and inventor, famous for his pioneering development of pocket calculators and home computers. He was a self-described libertarian who believed that Britain needed to be more open to those with talent and skills.[43] Among the organisations he supported was the Adam Smith Institute (ASI), Britain's foremost free market think tank of the era. The ASI, in turn, was founded and led by Madsen Pirie. Part of a transatlantic network of policy organisations that helped spread the gospel of free markets from the late 1970s, the ASI was particularly close to the government of Margaret Thatcher and helped to elaborate many of the policies – including privatisation, outsourcing and the Right to Buy scheme – which came to define the politics of the era.

Pirie thus played a key part in building the intellectual foundations of Thatcherism. But as his prominent role in Mensa would suggest, he also seems to have been very interested in the idea of intelligence. Alongside a shelf-full of weighty books on policy, philosophy and economics, he published a series of IQ quiz books co-written with one of the other founders of the ASI, Eamonn Butler. Titles like *Test your IQ, Boost your IQ* and *The Sherlock Holmes IQ Book* (subtitle: '*Test Your IQ Against the Great Detective*') tapped into the booming market for IQ-related self-help books in the 1980s and 1990s, which Mensa had helped to create. They also gave Pirie and Butler a platform to share their views on intelligence, including the idea that it was mainly inherited, and that you needed a certain IQ level to do certain jobs.[44]

Like Serebriakoff, Pirie was an ardent anti-socialist, telling an interviewer from the Mensa magazine that when he visited schools he liked to shock teachers by comparing the impact of socialism to that of civil war and genocide.[45] He also firmly believed that people, or at least intelligent people, made better choices than governments. In 1985 he regaled a Mensa gathering at Cambridge with his vision for a future where the 'talents and enterprise *of able people*' (my italics) had been released from all forms of oppressive government control.[46]

His views reflected a broader Mensa faith in meritocracy and social mobility – the idea that intelligent people from whatever background should be given the opportunity to rise to positions where they could have a positive impact on society. Serebriakoff, who was also a regular collaborator with the ASI, believed that this kind of social mobility of the able had played a key role in the transformations of the Industrial Revolution. The only way for societies to develop in the modern era, he argued, was to recreate those conditions by setting innovators and entrepreneurs free.[47] Reducing the power of the state, for men like Pirie, Sinclair and Serebriakoff, had the benefit not only of freeing markets, but also of freeing the most able in society to deploy their intelligence in whatever ways they saw fit.

Lisa Simpson's ambitions for the intelligent to improve society through the application of reason have certainly been shared by some of her real-life Mensa compatriots over the years. Like so much of the press coverage of Mensa, the show mocked these ideas and the apparent pretensions of Mensa members. But at the end of the 20th century, when politics had come to be defined by the technocratic vision of the Clinton and Blair era, the show also seemed to be mocking the pretensions of the self-appointed intellectual and educational elites, who increasingly dominated political life on both sides of the Atlantic. Its jokes landed partly because, by this point in time, the idea of rule by the intelligent had become familiar enough to be a target for satire.

Mensa has never been influential or important, and its popularity has always had hard limits. But the fact that it was able to emerge and thrive in the second half of the 20th century tells us that this was an era when ideas about intelligence were changing. From being a minority concern, intelligence had moved towards the social and cultural mainstream, more frequently referenced and increasingly valued. And with it went the idea that, in an increasingly complex world, societies would only be able to progress if the intelligent were allowed to rule.

6

Gifted and Talented

'... if the mass of man could clearly comprehend the true origin of advancement in civilization, doubtless gifted persons would be generally prized and explicitly rewarded'.

Leta Hollingworth

IN THE 1950s, AN IDEALISTIC young Italian priest was sent by the Church to work in the town of his birth – Petralia Soprana, high in the Madonie mountains of Sicily. Don Calogero La Placa had been born into a poor family, but had won a scholarship as a child, which had given him access to a world of education that otherwise would have been out of reach. During his seminary studies he had been supported by a bishop who recognised his academic talents. His community in Sicily was marred by poverty and isolation, a society only recently released from the shackles of feudalism. Most of its inhabitants scratched out a living by farming small plots of inhospitable mountainous land. Those who wanted to escape this life of hardship appeared to have only two options: emigrate or join the mafia.

Don Calogero spent his first years back in Petralia Soprana fighting to secure the community's access to the basic material needs of modern, civilised life: electricity supplies, clean water, sewage systems and so on. Once he had succeeded in upgrading these services, he turned to the second stage of his mission. From now on, his focus would be on what he described as 'upgrading the human being'.[1] To do this, he believed, the community must turn its attention to education. Although secondary education was technically compulsory in Italy, only a third of children in Sicily stayed in school after the age of 11. From that age, most children were needed to work on the family farm or find other ways to earn an income.[2] Don Calogero felt that the lack of education was holding Sicily back.

In the early 1960s, Don Calogero and his family began raising money to buy land for a new school. By 1967, supported by donations and bank loans – including a private donation from the President of Sicily – they had raised enough money to buy a plot of land. Perched on a plateau just outside Petralia Soprana, the 20-acre site was ringed by ice-capped mountains and overlooked an ancient Roman aqueduct nearby. At first it consisted of a single building – a dormitory, classroom, library, dining room and living space all rolled into one. But it soon expanded into what Don Calogero envisaged as a self-contained community, with separate dining rooms and dormitories, a library and a farm to produce its own food.

This, however, was not going to be a normal school. What Don Calogero wanted to build was a school for the *gifted*. As he travelled around the villages and hamlets of the region, he had been struck by the number of highly intelligent children who lacked the educational opportunities needed to nurture their talents. These were children, he believed, who had the potential to transform Sicilian society, to put their intelligence to the service of its cultural and economic development. But without the right opportunities their talents would be squandered, and these highly intelligent children would be lost into delinquency or a life of organised crime.

Don Calogero set out to find and support these children. He and his collaborators began by asking teachers across Sicily to refer their brightest and most creative pupils. Over 200 were referred. They were interviewed by social workers and given a battery of intelligence and aptitude tests by psychologists from the University of Palermo. Twenty of them were offered places at the school, all boys initially, although the plan was to open the school up to children of both sexes in time. By the second year there were 29 boys living and studying in the new school. It was the first school for gifted children in the whole of Italy.

Not everyone was happy about this idea. In a society marred by poverty, where rural children were often malnourished, many regarded the idea of a school for the gifted as frivolous. A lot of local families didn't trust priests as a point of principle. Some thought the focus on gifted children was elitist. Some didn't like the fact that all these resources were being given to the children of the poor. Others thought it ludicrous that such a project

Don Calogero. From 'Forging a New Society' (c. 1969), Nathaniel Weyl papers, box 49, folder 6, Hoover Institution Library & Archives.

could take on the mafia. 'Oh yeah,' scoffed an Italian-American man in New York asked to donate money to the school, 'those kids are smart and they'll become better crooks, better capo Mafiosi. Instead of stealing pennies, they'll steal dollars.'³

The school was designed to teach pupils between the ages of 11 and 18, preparing them for university life. It broadly followed the Italian public school curriculum, but with a very different educational approach. Like the other famous Italian educationalist, Maria Montessori, Don

Calogero believed that children should be given the freedom to develop their own interests and expertise. The school avoided formal discipline and rules, giving children the freedom to study as they wished. It aimed to encourage an atmosphere of curiosity and a genuine love of learning. School life revolved around open discussion, learning by doing, and arts, languages and music. Formal study was combined with work on the school's dairy farm and in its pizza restaurant. It enjoyed a high teacher-pupil ratio, with university-educated teachers coming to work there from countries around the world, from the United States to Nicaragua. Children were encouraged to learn about Sicily and its history, including the history of organised crime and corruption. They were encouraged to think of themselves as future leaders who could help transform Sicily into a prosperous modern society.

The new school attracted worldwide interest. The famous American anthropologist Margaret Mead visited and supported the school. 'This may well be,' she wrote in 1968, 'a unique experiment in salvaging high ability and developing indigenous leadership in a technologically backward and economically deprived country.'[4] She also sent a young American PhD student, Jo Danna – whose parents had emigrated to the US from Petralia Soprana before she was born – back to her ancestral homeland to study the cognitive abilities of local children and to support Don Calogero. The school attracted funding from philanthropic groups in Europe, was discussed by organisations like UNESCO and the US Educational Testing Service, and collaborated with experts from Jean Piaget's renowned Rousseau Institute in Geneva.

It was also closely tied to Mensa. Don Calogero had first seen an article about the organisation in an Italian newspaper in 1966, and had written to Victor Serebriakoff asking for help with his project. Invited to address Mensa gatherings, he denounced members for their frivolity and naval gazing. It was a crime, he argued, for them to waste their talents on a glorified social club. Instead, they should throw their weight behind projects like his, working to harness intelligence for the good of humanity.

It was this intervention that prompted Mensa to update its constitution, adding the commitment to 'identify and foster human intelligence for the benefit of humanity', which remains in place to this day. Mensa members from around the world campaigned for the school, donated money and travelled to Sicily to work as teachers. The school's first headteacher, Peter Edwards, was an English Mensa member who had recently graduated from Oxford University. He and other Mensans developed plans with Don Calogero to expand and internationalise the school, aiming eventually to turn it into an institution that would educate hundreds of children from Sicily, elsewhere in Italy and around the world.

These plans didn't bear fruit. The school survived the huge Sicilian earthquake of 1968, in part thanks to student volunteers who came from across Europe to help rebuild it. But despite raising considerable funds to cover the foundation and initial construction of the school, its backers were never able to secure enough to cover its annual running costs. It closed in 1975. Some of its former students went on to enjoy successful careers, including one as a surgeon in the United States. Don Calogero passed away in 2021 at the age of 96, and his death was marked by a wave of newspaper articles celebrating and remembering his school.

We've already seen how, in the 20th century, the idea took hold that progress in the modern world was dependent on a narrow, intelligent elite. The talented tenth, many came to believe, was the key to maintaining cultural standards, driving scientific innovation and defending the hierarchical social order against the egalitarian pressures of mass society.

For those who believed in the power and importance of a cognitive elite, the identification and training of that elite became an urgent political issue. This meant focusing on children. In societies around the world, officials, experts and educators began to ask how they could spot the most 'gifted' children and ensure that they were given the right education to prepare them for their role as future leaders. Don Calogero's school was a small part of a much bigger global movement.

So far we've focused our story of the history of intelligence in the Western, capitalist world. But much of the rest of the world for much of the 20th century was under communist rule. If we want to truly understand the modern history of intelligence we have to ask how it was understood in communist societies. One way to do so is to look at attitudes towards gifted children. Communist and capitalist countries had different ideas about intelligence and different approaches to education. But they were united by their interest in such children. On both sides of the ideological divide, governments and educators sought for ways to identify the most intelligent children and to forge them into a new elite.

Like Don Calogero's school, these project often failed, or at least failed to live up to the hopes that had been placed in them. It turned out that identifying the most gifted children, or even agreeing on what giftedness actually meant, was far from straightforward. Gifted education could be met with hostility from parents, teachers and politicians. The children themselves often failed to live up to the expectations set for them, or to contribute to society in the way their supporters had hoped. But faith in these children remained a constant across very different types of society all the way through the 20th century. As saw in Chapter 1, it remains a part of the global educational landscape today.

Clever communists

Vladimir Lenin died in January 1924 following a series of strokes and seizures. In the young Soviet republic he had been lionised for his unique revolutionary genius. His former colleagues in the Politburo were determined to prevent the secrets of this genius from dying with him. So they decided that his brain needed to be studied to find out more about what Lenin's genius consisted of and where it came from.[5]

The renowned German neuroscientist Oskar Vogt was recruited for the task. A pioneer of brain research, he had already established two brain institutes in Berlin and now came to do the same in Moscow. His work built on similar efforts – carried out over previous decades in Germany, Sweden and the US – to uncover the secrets of intelligence by studying the brains of gifted scientists and intellectuals. Many of those who studied 'elite brains' were also eugenicists who hoped that ways could be found to breed better brains in the human population. One of the first brains to be studied for this purpose was the German Carl Friedrich Gauss, who, among other things, was famous for discovering the 'bell curve' of normal distribution which was so important to the 20th-century science of IQ.

Vogt, his wife Cécile, and their team sliced up Lenin's brain into tens of thousands of sections, which were then mounted and stained for analysis. Their work formed the basis of a new brain research institute, which included a museum with a 'Pantheon of brains' from eminent Russians. After five years of research, Vogt announced that the neurons in part of Lenin's cerebral cortex were unusual in their size, shape and number, and that this would have made Lenin a 'mental athlete'. This, Vogt believed, explained the evidence from Lenin's friends and colleagues about his unique ability to think quickly, to observe carefully and to make sense of reality.[6] Over the following years Lenin's brain would be compared to other eminent brains held at the Moscow institute, including the novelist Maxim Gorky, the Nobel Prize-winning neurologist Ivan Pavlov and the pioneering German communist Clara Zetkin.[7]

The story of Lenin's brain suggests something that might seem surprising: that the leaders of the Soviet Union believed in the biological superiority of a small number of elite individuals. As Trotsky argued in 1925, geniuses such as Marx and Lenin only emerged once in a century, and 'no genius can be created even by the decree of the strongest and most disciplined party'.[8] We might expect Bolshevik leaders to have had a more egalitarian view of the world, to see progress as stemming from the masses, from ordinary workers and peasants. But their fascination with Lenin's brain seemed to suggest the opposite: that the progress of politics, culture and civilisation was driven from the top, by a biologically determined intellectual elite.

The concept of an intellectual elite – and the question of who should be part of it – was central to the revolution in Soviet education that took place during the 1920s.[9] When the Bolsheviks took power in 1917, they were met with deep hostility from much of the educated class. Tsarist education had been limited in size and suffused with class discrimination. Almost all of those with higher or technical education were from bourgeois or aristocratic backgrounds. There were almost no highly educated workers or peasants. Teachers went on strike against the Bolshevik regime in 1918, and the Bolsheviks enjoyed very little support from educated professionals like engineers and army officers. While the Soviet regime was fighting for survival during the civil war (1919–23) and its aftermath, it was forced to accept these bourgeois specialists whose skills were so desperately needed. But Soviet leaders believed that the long-term strength of the communist project could only be assured if they could build a new working-class intellectual elite, a politically loyal proletarian intelligentsia.[10]

Creating this new revolutionary class required a revolutionary approach to education. Educational reformers wanted to improve basic education and literacy levels for the whole of society. But their priority was developing a system of 'promotion' which would allow a new class of workers and peasants to access advanced education, acquiring the technical skills necessary to climb the professional ladder and replace the bourgeois specialists. This new educated elite would form the backbone of skilled workers, industrial managers and political cadres which would drive economic development and cement Soviet rule.

Marxist theory offered no blueprint for how to achieve this. Marx and Engels had written little about education. Neither had Lenin given the issue much thought before the revolution, although he had huge faith in the power of education and a fairly traditionalist view of the value of high culture and literature. A far more influential figure in the field was Lenin's wife, Nadezhda Krupskaya. Krupskaya was a leading Bolshevik activist from the pre-revolutionary era. After 1918, she helped shape Soviet education policy as one of leaders of the Commissariat of Enlightenment.

The transformation of education following the Bolshevik revolution was immense. The whole system was opened up and democratised. New progressive teaching methods were introduced in schools, replacing the Tsarist traditions of hierarchical, knowledge-based education with modern Western ideas about child-centred pedagogy. At first, all forms of school exams and formal assessment were scrapped, replaced by systems of self-assessment and verbal feedback focusing on things like creativity and skills.[11]

But there were different opinions about who should be educated and how. Some of the old Bolshevik intellectuals such as Krupskaya wanted to promote educational opportunities for all, regardless of background. Other groups, such as the Soviet youth movement the *Komsomol*, advocated more class-based affirmative action. Parallel systems of proletarian education were set up to give priority to workers and peasants – factory-based schools providing a route for young workers who had grown up without educational opportunities to gain a basic education and enter the universities. Access to higher education was thrown open, with the traditional requirement for a secondary school diploma abandoned. Routes to university came to depend far more on a person's class, background and political loyalty than on their educational credentials. This had a massive impact on the student body. By the early 1930s, workers represented more than half of all university students.[12]

This didn't mean, however, that Soviet educational reformers weren't interested in intelligence or the idea of an intelligent elite. Their goal wasn't just to radically raise the educational level of the entire working class. They also wanted to identify the brightest workers and peasants, and to channel educational resources their way to transform them into a proletarian intelligentsia. The system of proletarian education was part of this mission, designed to identify those with exceptional educational talents even if they had not been able to benefit from formal education prior to the revolution. There were even those who advocated for special schools and special welfare support for gifted children and geniuses.[13]

This focus on the 'brightest' was reflected in the popularity of intelligence testing during the 1920s and early 1930s.[14] Just like in other countries, Soviet psychologists and teachers saw these tests as a modern, scientific way to assess children. They offered an alternative to traditional academic exams that could be harnessed to the goal of moulding the next generation into new Soviet citizens. The use of intelligence tests was driven by experts working in the field of *pedology* – a discipline combining child psychology, education and health. Soviet pedologists translated versions of the Binet-Simon test and created their own intelligence tests. They studied the way such tests were being used in countries like the US and collaborated with intelligence testers abroad. Tests were widely used to stream children within Soviet schools during the 1920s and early 1930s.

Soviet society closely followed the global scientific debate about human genetics and 'nature versus nurture' that was taking place during the period. Early Soviet scientists were deeply engaged with the new science of genetics. But there was a widespread hope that scientists would prove that environmental factors were crucial in shaping human abilities. If the Soviet Union could, as its supporters hoped, transform the social environment its children were raised in, it would also be able to transform the potential of a new generation of Soviet citizens. Improved social conditions would make Soviet people more educatable, skilled and intelligent.

An example of this line of thought came from a young Soviet psychologist named Alexander Luria. At the beginning of the 1930s Luria travelled to Soviet Uzbekistan to carry out an innovative new study on mental development. Uzbek society had undergone a rapid transformation since the Bolshevik revolution. It had witnessed the beginnings of industrialisation, the introduction of new education systems and the relocation of previously nomadic populations to new collective farms. These changes, Luria believed, had fundamentally changed the ways in which people thought.

To prove this, Luria set out to test local people's conceptual and abstract thinking, and their ability to reason logically. He did so through the use of syllogisms – questions that involve logically inferring answers from certain premises – a standard technique for psychologists at the time. An example of one of his questions was as follows:

In the Far North, where there is snow, all bears are white. Novaya Zemlya is in the Far North and there is always snow there. What colour are the bears there?

The young people who had been living on the collective farms for a few years and had some experience of formal schooling, even if they remained illiterate, were able to answer this question easily. The bears, they understood, had to be white.

But when Luria put the same question to older workers who had grown up in remote communities with no formal instruction, he found that almost no one gave this answer. Indeed, most flat-out ignored the premise, simply refusing to make judgements about something they had no direct experience of, even when pressed to do so. In the words of one respondent: 'We always speak only of what we see; we don't talk about what we haven't seen . . . If a man was sixty or eighty and had seen a white bear and had told about it, he could be believed, but I've never seen one and hence I can't say. That's my last word!'[15]

For Luria, this showed that the transformation of Soviet society was being accompanied by a transformation of people's mental worlds. Older forms of social organisation limited the ways in which people could think about and understand the world. Modern societies, though, enabled people to think about the world in radically new ways. The changes to the way people perceived, deduced, reasoned, imagined and analysed were advancing human consciousness to new levels. Luria's arguments chimed with the Soviet officials and educationalists who believed that the new Soviet society would fundamentally transform people's mental abilities. Whether this is what was really going on with Luria's respondents is, of course, open to question.

In the first decade and a half of Soviet rule, a period that witnessed a series of dramatic political changes, attitudes to intelligence and the intellectually gifted were in flux. Soviet society valued high intelligence in many ways. The most intellectually gifted workers were understood as key to consolidating the revolution. And Soviet experts embraced parts of the new intelligence science emerging from the West. But when it came to education, a person's political loyalty and class background ultimately mattered more than their academic track record. And the key to progress was understood to lie as much in the intellectual transformation of the masses as in the abilities of a gifted vanguard.

We can see some of these ambiguities in the way that IQ tests were used. The pedologists who embedded intelligence testing into the Soviet school system were more concerned about academic *underperformance* than in the potential of the most gifted. Education leaders wanted to ensure a successful education for the mass of working-class and peasant children. They saw children with low mental levels as a barrier to this, creating disorder in the classroom and monopolising the attention of teachers. Identifying these children and removing them to special schools, they believed, would optimise the productive capacity of the new Soviet mass education machine.[16]

A big change in attitudes towards intelligence, intelligence testing and gifted children took place in the mid-1930s. At the heart of that change was a dramatic decree from the Politburo in July 1936 that banned the discipline of pedology altogether. This included a blanket ban on the use of all intelligence tests, and the dismantling of the system of school streaming and special schools for underperforming children. Intelligence testing would remain formally outlawed throughout the Soviet era.

The Soviet rejection of intelligence testing was sometimes interpreted in the West as evidence of greater Soviet sensitivity to the biases and limitations of intelligence science. This was partly true. The 1936 decree came in response to mounting public criticism of pedologists and their work. This criticism often focused on the disproportionally large number of proletarian, peasant and ethnic minority children being sent to special schools. Families of these children complained that they were being negatively labelled because of their class or racial background. As political debate over the problems in Soviet schools rose in the early 1930s, these criticisms were taken up by a number of high-profile officials. These officials criticised pedologists for pursuing a form of biosocial determinism that was biased against proletarian children. Newspapers began comparing pedologists to Nazis, arguing that they were promoting ideas about the biological superiority of ruling classes and races.[17]

But there were also other factors underpinning the 1936 decree. Some officials complained that the large number of children being sent to special schools reflected badly on the achievements of Soviet society. If so many children were being publicly labelled as intellectually backwards, they feared, it suggested that the Soviet system was failing in its mission to mould a new generation of citizens and to integrate all nationalities and classes into Soviet life. Testing and streaming, in other words, was bad publicity. Others alleged that pedologists were undermining the work of teachers. And some historians have argued that the decree was part of a wider Stalinist effort to undermine one of the last remaining power bases of the 'Old Bolsheviks' in the Commissariat for Enlightenment.

Whatever the motives, the 1936 decree symbolised a definitive break with the educational revolution of the early Soviet period. Like so many other areas of Soviet life, the experimentation and egalitarianism of early Bolshevik education ran aground against the Stalinist worship of tradition, hierarchy and authority. Out went all the egalitarian ideas about experimental pedagogy, open access to university, and special educational systems for workers and peasants. In came traditional teaching methods, a focus on knowledge, rigorous academic hierarchy, and rewards and social prestige for academic excellence. The traditional school exams and grading systems that had been abolished in 1918 were formally reinstated. By the late 1930s, social and political criteria for entry to higher education had largely been abandoned.

The 1936 decree against pedology and intelligence testing had explicitly denounced attempts to prove that certain children were specially gifted as a 'bourgeois idea'.[18] But this didn't mean that the Stalinist education system was no longer interested in the idea of an intellectual elite. What changed was the way that elite was defined and identified.

Attempts to separate out raw intellectual gifts from formal academic achievement were abandoned, whether in the form of intelligence testing or systems for workers to gain access to higher education without the right grades. They were replaced by a traditional reliance on performance in academic assessments. Now it didn't matter if you were a politically loyal worker denied access to education in your youth. If your grades didn't match those of the sons and daughters of the former aristocracy (or perhaps more relevantly the sons and daughters of the new breed of Soviet apparatchiks), your pathway to educational opportunity and social mobility was blocked. The aim of education was no longer to revolutionise society and expand human potential. Instead, its task was to identify a traditional academic elite that could be moulded into the high-end technicians, engineers and bureaucrats needed to drive industrialisation and manage the ship of state.

At the same time as this traditionalist turn was taking place in Soviet education, however, a new path emerged to identify gifted children outside of the formal school system. Olympiads in the maths and sciences became a huge cultural phenomenon from the middle of the 20th century, not only in the Soviet Union but in socialist states across Eastern Europe and beyond. Millions of children participated in them, and in later decades they came to be broadcast on nationwide radio and television. Winners of these competitions were granted places in elite universities, regardless of their formal qualifications. Many of the most prestigious mathematicians and scientists of the Soviet era began their careers as Olympiad winners.

Public maths and science competitions had their origins in the late 19th century. But the first organised Soviet competition was the Russian Maths Olympiad held in Leningrad in 1934.[19] It was organised by the country's leading mathematicians, and was expanded to Moscow and Kiev in 1935. This was not an easy time for the Soviet metropolitan intelligentsia, on the eve of the launch of Stalin's purges which would plunge the Soviet political and cultural elite into three years of mass terror, denunciation, exile and execution. But the Olympiads they organised not only survived; in the long run, they thrived.

Key to the Olympiads' success was the political thaw following the death of Stalin and Nikita Khrushchev's rise to power in the late 1950s. This was the peak of Soviet scientific self-confidence, symbolised by the launch of Sputnik – the first ever artificial satellite – which was interpreted around the world as proof of Soviet technical prowess and educational achievement. But it was also a moment that sparked an internal debate about scientific education. Now the Soviet Union had so clearly leapfrogged the West, future scientific advances were going to have to come from within rather than from emulating science produced abroad. Despite the controversy that continued to surround ideas about special education for the gifted, Khrushchev spoke frequently about the importance of developing the talents of Soviet youth.[20] In the early 1960s, a small number of science boarding schools was established, designed to educate the cream of young scientific talent. Alongside these were new special schools aimed at the most talented artists, athletes, linguists and even circus performers.

It was in this context that a call for the first all-Union scientific Olympiads was issued by the famous mathematician Mikhail Lavrentyev. Lavrentyev was the founder and leader of Akademgorodok, the secretive centre of Soviet research science established in the frozen Siberian wilderness in the late 1950s. Located over 3,000 kilometres from Moscow, the Soviet Academy of Sciences built an entire new town to bring together thousands of leading Soviet scientists, offering (relatively) luxurious living standards and a (relatively) free political environment, with the goal of supercharging a new era of Soviet scientific achievement.

In 1960, Lavrentyev published an article calling for a new system of national maths and science Olympiads. Such a system was needed, he argued, to identify the next generation of gifted young scientists who could fill the new science boarding schools and ultimately find their way to Akademgorodok. The need for such a system suggested that the turn back to traditional academic assessment and hierarchy that had taken place under Stalin was failing to identify the country's best intellectual talent. Gifted children were being left behind.

Lavrentyev's call was answered, and in 1963 a national Olympiad system was established. The system's statutes nodded towards a general goal of increasing public interest in maths and sciences, and to improving the overall quality of teaching. But its real priority was to identify the most gifted young Soviet people and to fast-track them into scientific careers.

The All-Union Olympiads ran from January to April each year. They began with first rounds held in individual schools, where problems were set and marked by teachers. They then progressed in stages up to the final round held in Moscow, where each region would send a team of their brightest students to spend a week solving problems set by leading scientists. The problems were designed to identify those with creativity and problem-solving skills, rather than just those with lots of knowledge who would be picked up by traditional exams. An elaborate preparation system was also established, with budding Olympians encouraged to build their skills in after-school clubs and study groups. There were particular efforts to identify talented students outside of the major cities with the most advanced school systems. The top students each year received prizes, diplomas and automatic entry to the most prestigious schools and university departments.[21]

The popularity of Olympiads was mirrored in other communist states after the Second World War. And it was in one of these states that the *international* Olympiad movement emerged. The first International Mathematical Olympiad took place in Romania in 1959, initially only involving participants from communist countries.[22] This would eventually develop into a system of international Olympiads across the scientific disciplines, which is still going strong today. The United States had begun to hold similar national competitions in the 1950s, but had refused invitations to attend international maths Olympiads behind the Iron Curtain until the 1970s.[23] In contrast, communist states, particularly the Soviet Union, set huge store by their performance in these competitions, promoting them as international measures of scientific progress and prestige. Soviet scientists almost always finished in the top five of the international physics Olympiad, winning first prize on over a dozen occasions.[24]

The history of the early Soviet education reforms, Stalin's prohibition of intelligence testing and the thriving culture of Olympiads suggested that the Soviet Union had a complicated attitude towards the idea of giftedness. This complexity was mirrored in other communist states. East German psychologists, for example, developed what they called a 'Marxist theory of giftedness'.[25] They denounced Western ideas about gifted children as a cynical ploy to justify the dominance of the bourgeoisie. Instead, they argued that being gifted was a matter of type rather than degree. Everyone had gifts, and everyone had the potential to develop their gifts to a high level. But different people were gifted in different areas, and the social environment people grew up in often limited their ability to develop their gifts to the fullest extent. As in the Soviet Union, they thought that a new socialist education system would transform human potential by finally offering everyone, particularly workers, the chance to fully develop their natural gifts. But also as in the Soviet Union, the East German education system sought to engender traditional academic success and focused resources on a narrow elite through special schools for the brightest students.[26]

As the study of Lenin's brain suggested, 20th-century communist societies were just as interested in intelligence as their Western counterparts. Education policies and attitudes towards gifted children may have varied over time, and sometimes seemed contradictory. But there remained a core belief that an elite of gifted children needed to be identified and educated to help drive social and scientific progress.

Canny capitalists

The West never really embraced the idea of Soviet-style Olympiads or elite schools for the best young scientists. What it had instead was the gifted education movement.

There had been some schemes in the United States to fast-track academically successful children through the school curriculum from the late 19th century. The first references to these children as 'gifted' began to emerge at the start of the 20th century, when there was a surge in popular interest in child prodigies. The new breed of tabloid newspaper latched on to the popularity of stories about precocious children, reporting on the exploits of chess prodigies, maths whizz-kids and pre-teen poets. Gifted children were transformed into celebrities.

But the beginning of gifted education as a *movement* was bound up with the origins of intelligence science. The man sometimes referred to as the 'father' of gifted education was Lewis Terman, who we encountered in Chapter 3. In 1921, he launched the first scientific study of gifted children, identifying high-IQ children from across California and tracking their development for the next 40 years. His project sparked huge popular interest in the idea of gifted education. Terman was inundated with letters from teachers and parents asking for advice about how to support their gifted children, and the project was publicised as far afield as South Africa, Malaysia and Australia.[27]

The gifted education movement, though, had more mothers than it did fathers. On the opposite side of the US to Terman, Leta Hollingworth played the biggest role in building gifted education into a discipline. Like Terman a psychologist with a teaching background, Hollingworth launched her own longitudinal study of gifted children in New York in the 1920s, going on to run a gifted education programme and publishing widely on the subject. Elsewhere, the scientific study of the gifted launched the careers of a number of women who would go on to become key figures within American psychology. In the UK, the National Association for Gifted Children was founded by a woman named Margaret Branch. Branch was a volunteer ambulance driver during the Spanish Civil War, a relief worker (and allegedly a spy) during the Second World War, and a social worker and psychotherapist in London after the war, where among other things she pioneered psychological care for the trans community and ran counselling services for LGBT charities throughout the AIDS crisis.

The work of people like Branch, Hollingworth and Terman helped to popularise the idea that the 'gifted' represented a distinct class of children who deserved a specially tailored form of education. In the interwar period, new schemes for gifted education proliferated across the US. Sometimes these involved educating gifted children in separate schools, but more often they focused on providing an enriched curriculum for gifted children within the regular school system, either through extracurricular classes or internal streaming.

Gifted education, as we saw with the story of Don Calogero's school at the start of this chapter, was always somewhat controversial. Right from the start of the 20th century, critics labelled it as elitist, focusing resources on a minority of students with the least urgent needs at the expense of the majority of the 'non-gifted'. Despite these criticisms, the general trend in gifted education has been upwards. In the 1930s, for example, there were fewer than 2,000 children enrolled in official gifted education programmes across the US. By the 1990s there were almost 2 million.[28]

Who exactly were these 'gifted' children? The answer wasn't obvious. They were very much a 20th-century creation, and those who promoted the idea had no historically accepted definition or measure to fall back on. But definitions were important; if the gifted couldn't be objectively identified, they would remain just an interesting set of individuals rather than a distinct category of children with their own set of requirements.

In the early 20th century, those who promoted gifted education focused almost entirely on the new measure of IQ to identify the gifted, reflecting the fact that the study of gifted children emerged directly from the new science of intelligence. For those who relied on IQ, the only real disagreement lay in where the cutoff point was. Was anyone with an IQ over 120 gifted? Or should the limit be set at 140 or 160?

Even from the 1930s there were those who criticised IQ tests as far too limited a way of capturing the diversity of children's gifts. As the century wore on, gifted education practitioners gradually turned away from using IQ tests alone to identify gifted children, combining them with information from teachers, parents and other groups. They also increasingly began to talk about 'gifts' in the plural, looking beyond a narrow definition of intellectual giftedness to focus on qualities like creativity.

Those who pioneered the category of gifted children wanted to do more than just define them. They also wanted to promote those children to the world, to reveal the full scope of their abilities, to demonstrate how their gifts could be harnessed to transform society. As Terman argued as early as 1920, 'whether civilization continues to advance depends very largely on the contributions made by creative thinkers and leaders . . . The average man can follow when genius has shown the way. The genius of tomorrow is the gifted child of today.'[29]

Part of the difficulty in promoting gifted education was the need to challenge negative stereotypes about intellectual high-achievers. These stereotypes had partly emerged from the popular association between genius and madness that had taken root in the public mind during the 19th century. But many people at the start of the 20th century assumed that very clever children must be deficient in other ways – physically weak, unhealthy and lacking in common sense and social skills. Nerds, in other words.

So the gifted children pioneers like Hollingworth and Terman set out to show that such children were not just intellectually better than the rest – they were also morally, physically and spiritually superior. They collected data to show that gifted children were better behaved, more truthful, more athletic, more musical and more emotionally stable. Leta Hollingworth set up a special study to prove that gifted children were more beautiful than ordinary children. Lewis Terman, with the unerring ability of 20th-century intelligence scientists to make things weird, claimed that gifted children developed pubic hair at an earlier age than the non-gifted.[30]

There was a problem, though, with this line of argument. If gifted children were so well-rounded and successful, how could anyone justify channelling more resources towards developing a special kind of education just for them? The answer, for gifted education advocates, was to highlight their vulnerability to the non-gifted.[31] The non-gifted, people like Terman and Hollingworth argued, were scornful towards the gifted. They were envious. They bullied and exploited them. And they were too unsophisticated to appreciate the complex needs of the gifted child. Left at their mercy, without special support from the education system, the potential of gifted children would be smothered by non-gifted resentment.

This mattered, the supporters of gifted education claimed, because gifted children were valuable to society. For one thing, they were society's future leaders. No one, Hollingworth argued, wanted a stupid person as a leader. Terman worked hard to demonstrate the achievements of gifted children as they grew older, tracking how many appeared in *Who's Who*, and how many patents, books and scientific papers they produced. In the second half of the 20th century, gifted education frequently focused on instilling children's sense of self-worth, their social confidence and their perception of themselves as part of a class whose destiny it was to direct and lead the non-gifted. As one scholar of the movement described it, gifted education became a 'pedagogy of the privileged'.[32]

Such future leaders, advocates hoped, could help tackle some of the biggest problems facing the modern world. As Jennifer Crane – a leading historian of gifted education – has shown, gifted children were often understood as vehicles for the promotion of democracy, peace and international co-operation. Gifted children in Britain during the 1970s were taught Esperanto so they could better contribute to the mission of world understanding.[33] After the fall of communism in Eastern Europe, gifted education experts argued that the talents of gifted European youth could be used to reunite the continent and secure lasting peace.[34]

But it was more common to frame the value of gifted children in national terms, as natural resources nations could exploit to achieve wealth and progress. Sometimes this value was seen as purely economic. 'The world of work is arranged in a hierarchy with reference to degree of intellect,' Leta Hollingworth wrote in 1936. 'The top one per cent can do all that the rest can do and some things that none of the rest can do.'[35] Cultivating the skills of this 1 per cent, advocates argued, could boost a nation's human capital, driving national growth and productivity.

More often, though, the value of gifted children to the nation was framed in military terms. This was especially true during the Cold War. Western states were thrown into a panic in the 1950s about the scientific and technological progress of the Soviet Union. Desperately seeking explanations for how the Soviets had apparently leapfrogged the capitalist world, many commentators looked at things like the science and maths Olympiads and concluded that the secret of Soviet success lay in its ability to identify and cultivate a talented scientific elite. High-profile figures like Admiral Hyman G. Rickover, the father of the US nuclear submarine programme, argued that the West needed to unshackle the talents of its gifted children and harness them to the global fight against communism. Arguments like these led to the 1958 National Defence Education Act, which sought to improve maths and science education for the brightest students. Similar ideas emerged when Cold War tensions rose again in the early 1980s. In 1983, Ronald Reagan's Department of Education published a report, *A Nation at Risk*, which claimed that a national decline in educational standards since the 1960s represented 'an act of unthinking, unilateral educational disarmament'.[36]

But beneath this boosterish rhetoric there were major controversies over the ways in which gifted education was being implemented on the ground. In particular, the organisation of gifted education programmes prompted concerns about the biases that were embedded in the identification of the gifted. Gifted education was sold as being democratic and meritocratic, providing equal opportunities to talented children regardless of their background. Perhaps unsurprisingly, the reality tended to fall short of this rhetoric.

These issues were evident right from the start of Terman's first study of gifted children in the 1920s. His method of taking recommendations from teachers and then assessing children against his culturally biased new intelligence tests meant that his test group was dominated by White, middle-class boys. Terman argued that the underrepresentation of girls was due to the fact that boys had greater IQ variability.[37] In reality, the discrepancy was almost certainly due to differences in referrals by teachers, who tended to judge boys' intelligence more highly than girls'. Later gifted child studies largely rejected Terman's arguments about intelligence variability, but gender biases remained embedded in the field. Well into the 1980s, examples of young girls speaking eloquently about their desire to become teachers, nurses, artists or mothers were being used to illustrate their giftedness, while gifted boys were praised for voicing their ambitions to become scientists, lawyers or politicians.[38]

Terman's work also showcased the racial biases that would characterise the gifted education movement throughout the 20th century. His gifted children study group was primarily made up of White children of English, Scottish and Jewish parents, with very few children from Mexican, Asian, southern European or Black families. Rather than prompting questions about his methodology, Terman used this fact to support his argument that White western Europeans and Jews were disproportionately gifted. Leta Hollingworth also argued that there were few gifted Black children.[39] These arguments were embedded in contemporary debates about race, intelligence and 'nature vs nurture'. The popular interest in geniuses and gifted children in the early 20th century reflected the strength of Social Darwinism at the time. Terman and Hollingworth, like many liberal psychologists and progressive educational reformers of the era, were active in the American eugenics movement.

Overtly racist language largely disappeared from the gifted education movement by the 1960s. But that doesn't mean that racist ideas about giftedness had gone away. According to Leslie Margolin, one of the most perceptive scholars of the movement, the growing American interest in gifted education had more to do with race and racism than with the Cold War and Sputnik.[40] As the proportion of non-White children in American schools increased and desegregation was promoted by the federal government, gifted education programmes offered a silent mechanism through which predominantly White, middle-class children could be streamed and separated within the 'overcrowded' school system, preserving their access to a superior education. Many within the gifted education movement, on both sides of the Atlantic, continued to assume that Black and working-class families couldn't provide nourishing environments for their gifted children. The only way such children could fulfil their potential, they argued, was to study and socialise with other gifted children, assimilating with the White, middle-class culture embodied in gifted schools and clubs.[41]

The problems of these class- and race-based exclusions were increasingly recognised by gifted educational experts from the 1970s. The movement embraced the language of diversity, opportunity and inclusion, searching for new ways to remove bias from the system. But Black children in particular continued to be disproportionately under-identified as gifted. By the early 1990s, 5.4 per cent of all White children in America were enrolled in gifted education programmes, compared to just 2.4 per cent of Blacks and Hispanics.[42]

In fact, if we're looking for a genuinely meritocratic, diverse celebration of intellectually gifted children in Western society – and perhaps the nearest equivalent to the spectacle of the Soviet Olympiads – we have to turn to the grand tradition of the American spelling bee. Although it might seem frivolous compared to the laser-like Soviet focus on advanced maths and physics, the peculiarly fiendish nature of English spelling has meant it came to be seen as a test of intellect. American culture in particular has always been peculiarly fixated on the idea that a wide vocabulary is a mark of intellectual prestige.

Like the Olympiads, the modern spelling bee (a 'bee' was a term used for social events in early American colonial society) built on traditions stretching back beyond the 20th century.[43] Spelling bees had been a feature of colonial education in 18th-century New England, and had developed into a popular nationwide social activity in the 19th century. Their popularity spread further with the advent of modern media. In 1930, national radio broadcast a spelling competition between congressmen and news reporters in Washington DC as primetime entertainment.

One of the attractions of spelling bees was that they were seen as genuinely meritocratic. Spelling excellence wasn't the preserve of the highly educated elite, but could be found among every class, race and gender. Spelling bees could thus be incorporated into the myth of the American melting pot, where everyone had equal opportunities to succeed whatever their identity or background. Indeed, the first ever national spelling competition, held in Cleveland in 1908, was won by a 14-year-old Black girl named Marie Golden.

The modern version of the American spelling bee formally began in 1925, launched by a newspaper in Louisville,

Kentucky, which encouraged other newspapers around the country to sponsor winners from each of their regions. A staggering *2 million* children competed in local competitions around the country. Nine boys and girls were selected for the final, held in Washington DC. There they were invited to the White House to meet the president, Calvin Coolidge, beginning a tradition that continues to this day. The winner received a gold medal and a cash prize of $500 (about $9,000 in today's money).

The competition has grown ever since, with more and more local participants, bigger and more elaborate final events, and ever-greater amounts of prize money. Winners have traditionally been lauded in the local and national media as the smartest kids in the nation, regularly invited to appear on primetime TV from the *Ed Sullivan Show* to *Good Morning America*. From 1994 the finals were broadcast live on ESPN, further magnifying public attention and catapulting winners to new heights of fame.

Throughout this time, public focus has remained on the diversity of spelling bee winners. Unlike the Soviet Olympiads, which remained dominated by boys, there has always been a gender balance when it comes to spelling bee winners. Champions have emerged from small towns with one-room schoolhouses and from inner-city public schools. And spelling bee fans have always emphasised the success of immigrant children from every conceivable national and racial background, particularly celebrating stories of those who arrived in America as young children with no English but who in a few short years managed to reach the elite of spelling mastery. The spelling bee seemed to be the one area of modern American society where the only criteria for success was hard work and raw brainpower.

From American spelling bees to Soviet Olympiads, gifted children have been celebrated in popular culture and prioritised in national education systems since the start of the 20th century. Those who promoted the idea of special education for the gifted appealed to ideas of fairness and equality: the most intelligent children should be given the opportunity to develop their talents to the fullest, whatever their backgrounds, and should not be ignored or deprioritised in favour of those with more obvious needs. They also appealed to the idea of progress: the most gifted children, properly supported, would grow up with the skills and abilities to lead society, to advance science, to develop the economy, and thus to improve the lives of everyone.

There was little, it seemed, to object to in these arguments. Yet criticisms of gifted education persisted. It was hard to shake the accusation that ideas of giftedness were plagued by biases of race, gender and class; that they were underpinned by a hierarchical notion of human intelligence and value; and that spending more on the education of the gifted inevitably meant spending less on everyone else.

Even some of those labelled as intelligent felt ambivalent about the label and the advantages that were attached to it. Andrée Blouin, for example, was one of the most important figures in the 20th-century fight against colonialism in Africa. Born in the early 1920s in what is now the Central African Republic, she helped fight for independence in Guinea and in Congo, where she was one of the leading advisors to Patrice Lumumba. The European press labelled her the 'Black Pasionaria', and the Belgians and their Congolese proxies were terrified of her power and influence.

Blouin was the daughter of a White father and a Black mother. Her relationship with her father was complicated, at a time when European men living in colonial Africa rarely acknowledged their mixed-race children. When she was in her early 20s, her father legally recognised her as his daughter. As he was doing so, he explained the reason behind his decision: 'If you had not turned out to be such an intelligent child,' he told her, 'I would not have recognised you.'

At that moment, Blouin was being conferred certain privileges on the grounds of her intelligence. In a society where the intellectual superiority of White over Black was taken as a given, her giftedness was being used to justify her elevation to legal and economic rights that otherwise would have been denied her. Blouin, however, was not willing to accept such an argument. 'I answered nothing when he said those words,' she later wrote. 'Even if I had not been a child who tried to raise herself in the world, I would not have been less his child for that. Perhaps he thought it would give me satisfaction to think, "Ah, my father finds me intelligent." But that did not matter to me.'[44]

The intelligence of gifted children came to matter a lot to many people during the 20th century. But there were plenty of others who refused to accept that it really mattered at all.

7

Building an Artificial Brain

'How will it end? I suggest the simplest way to find out is to make the thing and see.'

Ross Ashby

IN THE AFTERMATH OF the Second World War the eyes of the world turned, briefly, to Barnwood House, a mental hospital on the outskirts of Gloucester in the west of England. For years, the institution's chief psychiatrist, Ross Ashby, had been retiring to his laboratory at the end of the day to work on his scientific side hustle. Scavenging discarded equipment from army surplus, he had been building a mysterious new machine which he called the 'homeostat'. The machine consisted of a bulky collection of black metal boxes containing magnets and electrical circuits. No one knew what it was for. All Ashby would reveal was that it was designed to 'return to equilibrium in response to disruption from its environment'. By the winter of 1948 he was ready to reveal it to the world.

The world was, perhaps surprisingly, wowed. The reason such a strange contraption prompted so much excitement

W. Ross Ashby's 1948 Homeostat. Source: Wikimedia Commons

was Ashby's claim that, by using feedback to learn how to adapt to its environment, the homeostat was *thinking*. It was, to all intents and purposes, an artificial brain.

The news sparked excitement around the globe, appearing in press reports from Canada to Colombia. *Time* magazine in the US stated that it was 'the closest thing to a synthetic human brain so far designed by man'.[1] If scaled up, the *Daily Mail* reported excitedly, the machine could be taught to play chess.[2] A newspaper in India even claimed that the artificial brain had the potential to 'solve the world's economic and political problems'.[3]

Very few people today have heard of Ashby or his homeostat machine, and the idea that a small, self-correcting circuit of magnets could really be thought of as an artificial brain seems somewhat quaint. But the idea that machines can *think*, that we can create something that is able to learn and adapt independently of human instructions, is something we're much more familiar with.

Today we're living through what looks like the start of a new age of artificial intelligence, one which many people think is going to change the world in momentous ways. The idea of building machines that can think has a very long history. But today's AI age was born in the period between the 1930s and the 1950s, when a generation of mathematicians, engineers and neuroscientists began in earnest to pursue the idea of building an artificial brain.

The 'brains' they ended up creating were the first modern computers. We tend to think of computers as something that preceded AI – that computers were invented first, and people then began wondering whether they could become truly intelligent. But this is a misnomer; the two things were always understood as part of the same project. The people who invented computers did so in the belief that they would be able to think like humans do.

These people were, perhaps not surprisingly, deeply interested in the nature of human intelligence. But that doesn't mean they were obsessed with IQ and intelligence testing. Their interest lay in human intelligence *in general*, rather than the idea that some humans are more intelligent than others. As a result, they were much more interested in neuroscience and the function of the human brain than in the science of IQ.

Their views on human intelligence certainly displayed some biases and blind spots. As a rule, they were interested in the most obviously rational functions of the human brain – reasoning and problem-solving, games and puzzles. In this sense, at least, they resembled some of the early 20th-century intelligence scientists who had tried to boil down the mysteries of the human mind to a handful of mechanistic functions. And there certainly were individuals involved in the birth of the artificial brain who *were* very interested in different levels of human intelligence and what that meant for society – a dark side of the computing revolution which would resurface again and again over the coming decades.

The birth of the computer launched a revolution that has transformed our economy, our culture and our everyday lives. It was rooted in the quest to build a thinking machine – an artificial brain – which would ultimately be able to demonstrate an artificial version of human intelligence. This quest was built on the history of ideas about human intelligence, and in turn changed the way we have thought about intelligence ever since.

The deep history of AI

For thousands of years, humans have been imagining what it would be like if machines could think.

In Homer's *Iliad*, the god Hephaestus created female servants made of gold and possessed of understanding and speech. In the Middle Ages, Arab, Persian and European scholars, particularly those interested in alchemy, wrote instruction manuals for creating a small, artificial human called a 'homunculus'. Stories later emerged of philosophers such as Albertus Magnus and Roger Bacon creating model human heads that could speak, think and predict the future. Similar stories were told about the Golem, an artificial creature created by rabbis in medieval Europe to serve and protect Jewish communities.[4]

Alongside this history of ideas about intelligent machines, there is an equally long history of people building machines to help humans think better. The abacus was used in ancient China and Greece to assist humans with the types of calculations that couldn't be done on fingers and toes. From the 17th century, the slide rule helped engineers and architects carry out much more complex calculations. Around the same time, philosophers such as Leibniz and Pascal started to design the first mechanical calculating machines.

The first modern computers, or at least the first machines trying to do the kind of things modern computers do, were designed by the English mathematician Charles Babbage in the early 19th century. His initial design, the so-called Difference Engine, was an artificial calculator which used gear wheels to calculate mathematical tables. He followed this with the Analytical Engine, which was more akin to the general-purpose computer which would emerge in the 20th century. Designed to be steam powered, it was to be programmed using punch cards of the sort used at the time in Jacquard looms. It included a system for storing and accessing 'memory', alongside a central processing unit, and was able to modify itself midway through calculations.

Both of Babbage's designs were mathematically feasible, but they pushed the engineering techniques of the era to their absolute limits. A small prototype Difference Engine was eventually built after vast struggle and expense. But a working Analytical Engine proved a step too far. Indeed, it was not until the 1990s that a team at London's Science Museum was able to demonstrate the accuracy of Babbage's design by producing a fully working model.

It was Ada Lovelace, the English mathematician who collaborated with Babbage on the Analytical Engine, who was the first to explore the true intellectual potential of such a machine. Lovelace was the daughter of the notorious poet, Lord Byron. Her mother, Lady Byron, was a talented mathematician in her own right. Fearful that Ada would inherit her father's destructive artistic temperament, she ensured that her daughter's education revolved around maths, mechanics and logic.

Over recent decades, Lovelace's technical analysis of the Analytical Engine and the mathematics underpinning it has come to be recognised as a foundational moment in the history of computer science. Indeed, she was arguably the first ever computer scientist. And thanks to her reflections on the power and potential of the Analytical Engine – her comparison between the machine and the human brain – she is also understood today as one of the pioneers of AI.

In her famous commentary on the Analytical Engine published in 1843, Lovelace set out her vision of what a computing machine could achieve. She was careful not to exaggerate. The Analytical Engine, she argued, could not 'originate' anything, and could 'only do whatever we *know how to order it* to perform'. But that didn't mean that what it could achieve was restricted to only what already existed in human minds. It could be used, she argued, 'for determining that which human brains find it difficult or impossible to work out'. It could solve problems without a human having to work out in advance the precise formula needed. It could, she suggested, produce results which 'we might not otherwise have thought of obtaining'. With the right understanding of sound it could even 'compose elaborate and scientific pieces of music'.[5] Some 180 years later, her predictions seem eerily prescient.

As engineering advanced, the creation of more complex versions of these early computing machines became possible. The end of the 19th century saw the development of a new generation of punch-card calculating machines. The increasing complexity of government and administration was key to driving demand. In 1890, for example, the US census collected more data than human workers could analyse. To address this problem, the Census Bureau commissioned a specially designed electromechanical punch-card machine to tabulate the data. The company that made the machine merged with others after the First

World War to form a new company – International Business Machines, or IBM. By the start of the Second World War, IBM engineers had built on these foundations to develop much larger and more ambitious electromechanical calculators operated by circuits of electrically operated mechanical switches called relays, which the press had already begun to christen 'automatic brains'.[6]

These new machines were deeply impressive. But they were soon to be eclipsed by the machines that would come to define the next stage of the computing revolution – electrical, binary, general-purpose computers. For the first time in history, the world was presented with something that seemed to genuinely resemble an artificial human brain.

The birth of computers

Computers weren't invented by a single person. Like most modern technologies they were developed in stages by lots of different groups operating in different parts of the world at roughly the same time. But if there's one person we can identify as central to the development of computing and AI, it's Alan Turing.

Turing's work at Bletchley Park during the Second World War helped to crack the German Enigma machine, giving access to encrypted German messages and making an incalculable contribution to the Allied war effort. The techniques and machines he invented to do this helped to lay the foundations for the development of modern computers. But his life was marred by the persecution he faced over his homosexuality, which was illegal at the time. In 1952 he was prosecuted for homosexual acts and forced to undergo a form of chemical castration. He died at home two years later at the age of 41 from cyanide poisoning. Whether he was driven to suicide by the persecution he faced or was the victim of a tragic accident, we still don't know.

If you've seen the movie *The Imitation Game* you'll have some idea of who Turing was and what he achieved. The way Turing is introduced to us in the movie, however, is a bit misleading. The first impression we have of him is of an awkward, bumbling figure meeting with the military top brass at the start of the war and mumbling something about being good at puzzles. Turing was indeed an eccentric and sometimes awkward individual. Friends of his at Bletchley described how he would cycle to work in a gas mask to protect himself from pollen, and would chain his tin mug to the radiator to stop it being stolen. They regarded these eccentricities as a product of Turing's naïve, almost childlike simplicity, a simplicity that delighted his friends but which often embarrassed or infuriated other colleagues.[7]

But Turing didn't just turn up at Bletchley as someone who was good at puzzles. Even before his exploits during the war, he was already widely recognised as one of the most important mathematicians of his generation. The reason for his professional prestige was a paper he published in 1936 with the headline-grabbing title 'On Computable Numbers, with an Application to the *Entscheidungsproblem*'. This was also the paper that, many people argue, launched the 20th-century field of computer science.

The *Entscheidungsproblem* was a major challenge in the field at the time, one which effectively asked whether there were any mathematical problems that were fundamentally unsolvable. Turing's paper proved that there were. His proof rested on the use of a computing machine, which came to be known as a 'Turing machine'. Crucially, he was able to prove there were certain problems that would always remain logically impossible for the machine to solve.[8]

The Turing machine was not, it needs be emphasised, an *actual* machine. Rather, it was a theoretical machine – an abstract tool for thinking with – the design of which Turing sketched out in his paper. This machine differed from existing computing machines, which had so far only been designed to carry out specific calculations. By using stored programmes, the Turing machine would theoretically be able to do *any kind* of calculation by breaking it down into a sequence of component steps.

Turing's solution of the *Entscheidungsproblem* established him, still only in his mid-20s, as one of the leading mathematicians of his age. And his 'Turing machine' offered a blueprint for the development of a completely new type of computer. But, like Babbage and Lovelace a century earlier, his ideas had outstripped the engineering limits of his day.

All that changed with the outbreak of the Second World War three years later. The war was absolutely crucial to the development of modern computers and computer science. On both sides of the conflict scientists of all stripes were drafted into the war effort, asked to bring their expertise to bear on the burning military questions of the day. This sudden laser focus of resources and expertise on a narrow set of problems, often bringing together experts from fields who usually had little contact with each other, turbocharged both the demand for computing machines and ideas about how to build them.

In Germany and the US, mathematicians and engineers developed plans for new computers to support the development of guided missiles, to help automatically guide anti-aircraft guns, and to calculate artillery firing tables or the shockwaves of explosives. In Britain, the focus was on code-breaking.

Turing's work at Bletchley Park during the war is most famous for the electromechanical 'Bombe' machine that cracked Enigma. But from the point of view of computer history, the most important machines to come out of Bletchley were the Colossus computers. These were designed later in the war to decrypt the even more advanced German wireless teleprinter system nicknamed 'Tunny'. And although Turing helped with the design, the development of Colossus I and II was led by an engineer in the General Post Office named Tommy Flowers. Flowers' machines used vacuum tubes rather than the much slower relays, and were the first ever large, electronic, digital computers.[9]

Most of the Colossus computers were destroyed at the end of the war. They remained official secrets for decades to come. But the breakneck developments of the war years birthed a new generation of computers in its immediate aftermath.

In the United States, the mathematician John von Neumann (now perhaps best known as the inventor of game theory) had spent the war helping the military develop a universal electronic computer called ENIAC. In 1945 he drafted a report detailing how to construct something that resembled a real-life Turing machine – a stored-program electronic computer that treated program instructions the same way it treated data, doing away with the need for manual programming with switchboards or punch cards.

Armed with these instructions, engineers began to build what were in effect the first prototypes of modern electronic computers. At Manchester University, Turing's new team built the Manchester Baby and Manchester Mark 1 machines, the first operational stored-program computers, between 1948 and 1949. These were followed by other models in the UK, Australia and the US. At the Institute of Advanced Study at Princeton, von Neumann was working to build a much bigger computer which could help with the massive calculations needed to develop the first hydrogen bomb. By 1953 IBM had begun to commercialise these new devices, marking the launch of the modern computer industry.

Artificial brains

Thus was the modern computer born. At this stage a computer may have been roughly the size of your living room and cost the GDP of a small country. And what it could do, while incredibly impressive by the standards of the time, pales in comparison to the simplest computer we might come across today. But despite these limitations, those who had helped to invent the computer were convinced that they had succeeded in building an artificial brain.

This was particularly true of Alan Turing. When the Second World War ended in 1945, he had immediately turned his attention to the task of building on the achievements of the war to create a real-life version of the Turing machine. He told colleagues he was basing his designs around models of the functioning of the human brain. He described the analogy between computer and brain as the 'guiding principle' of his work.[10]

Turing's idea of designing computers' architecture on the model of the human brain showed how important neuroscience was to the new field of computing. The Greek physician Galen had pioneered the study of the brain and nervous system in ancient Rome, and his work had been developed by Arab and European scholars in the Middle Ages. But the first detailed study of brain function wasn't carried out until the 17th century, and neuroscience as a modern discipline didn't begin to emerge until the 19th century.

Scientists in this new field first began by exploring the areas of the brain and their various functions, identifying the frontal lobe as the site of intelligence. Towards the end of the 19th century they found the first evidence of electrical activity in the brain. The Spanish scientist

Santiago Ramón y Cajal helped to uncover the structure and function of brain cells at the end of the 19th century. By the start of the 20th century, neuroscientists had developed a model of the brain composed of billions of neurons communicating with each other via electrical impulses travelling through synapses.[11]

The vital role of electricity in the brain piqued the interest of electrical engineers. The fibres of nerve cells could exist in two states – either carrying a message via electrical impulses, or not. In that regard they looked very much like computer valves, whether electromechanical relay switches or electronic vacuum tubes, which were also governed by electrical impulses and could only ever be in one of two states. The two possible values of a nerve fibre or a computer valve could be represented in a binary form as 0 or 1. In the 19th century the English mathematician George Boole had introduced a new branch of algebra which, rather than using numbers and arithmetic, was limited to two values, denoted with 1 and 0, and some basic logical relations. In the 1930s, an American electrical engineer named Claude Shannon worked out how to apply Boolean logic to circuits, showing how binary electrical circuits could be used to perform complex mathematics.

A circuit integrating electrical valves with logic gates using Boolean algebra could thus, in theory, function in the same way as a human brain made up of networks of neurons and synapses. This was the idea that lay at the heart of the project to build a computer that would function like a human brain.

It wasn't just Turing and the electrical engineers who were interested in this project. It also inspired scientists working in a revolutionary new field which encompassed people like Ross Ashby, the inventor of the homeostat machine which we saw at the start of this chapter. It is a field that has now largely been forgotten, but it captured the imagination of some of the world's leading scientists and thinkers in the middle of the 20th century. They called it cybernetics.

The person who coined the term 'cybernetics' and was most associated with the field was the American mathematician Norbert Wiener. Wiener was a former child prodigy who enjoyed a meteoric academic career, starting college at age 11 and receiving a philosophy degree from Harvard at the age of 18. He spent almost his entire career at MIT, ending up as a lauded but eccentric figure who would roam the campus every day talking to himself, debating with his students, or barging into his colleagues' labs demanding to discuss whatever it was they were working on.

Like Turing, he also had a difficult personal life. He was emotionally volatile and prone to depressive and suicidal episodes. He was of Jewish descent and married a German woman, Margaret, who became increasingly pro-Nazi during the 1930s, to such an extent that she took to keeping a copy of *Mein Kampf* on her bedside table. After the Second World War she told Wiener that his two closest professional collaborators had slept with their daughter (they hadn't), causing him to break off all contact with them without ever explaining why he had done so. One of those colleagues ended up drinking himself to death.[12]

Cybernetics was a complex field bringing together scientists from various disciplines doing very different types of work. At its heart lay the study of control and communication – how animals and machines processed and communicated information, and how this shaped the ways they functioned or behaved. Wiener based the name of the discipline on the Greek word for a steersman – *kubernetes* – a figure who took in information about their environment and used that information to control the direction of a vessel. Within this broad definition there were a few big issues that attracted particular interest from cyberneticians: the ways machines and animals used feedback to modify their behaviour; the ways these machines and animals were affected by, and adapted to, their environments; and the ways in which humans and machines interact with each other.

The field grew out of ideas about the relationship between humans and machines in the interwar period.[13] But, as with the development of computers and computer science, it received a huge boost from the Second World War. Wiener, for example, was drafted into projects on predicting the flight of enemy aeroplanes and improving the accuracy of anti-aircraft fire, work which helped him to develop his theories about feedback and control. The discipline began to crystallise around 1942, when Wiener started organising a series of seminars bringing together mathematicians, engineers, neuroscientists and even anthropologists like Margaret Mead. In 1948 he published his most famous book, *Cybernetics: Or Control and*

Communication in the Animal and Machine, which first brought the discipline to wider attention. It sparked a huge wave of scientific and public interest during the 1950s.

Cybernetic ideas influenced new military technologies and systems in the US, from mechanical arms to handle radioactive material to the military's development of an 11-foot-tall walking truck.[14] It also inspired some of the era's pioneering work on DNA and biological systems, and in the emerging fields of AI and cognitive science. And it enjoyed global popularity. In India, Prime Minister Jawaharlal Nehru thought cybernetics could help the country to develop automated factories, and set up a system of institutes to train the technicians to run them. In the 1960s, the field became particularly popular in the Soviet Union, which developed ambitious plans to integrate cybernetics into education and industry, and to develop its own computing industry and internet.[15] In the early 1970s, Salvador Allende's socialist government in Chile developed a computer-based system to control the Chilean economy, Project Cybersyn, designed by the British cybernetician Stafford Beer.[16]

The popularity of cybernetics in the Soviet Union and other socialist states contributed to its decline in the West from the 1970s, particularly as US military funding was channelled into other fields. Its history is now largely forgotten, although its legacy lives on in various ways – from the 'cyber' in cyborgs and cyberpunk to the man-machine military hybrids in the *Iron Man* comics and movies.[17]

Cyberneticians' interest in information, feedback and the relationship between machines and humans meant that they were deeply involved with the development of computing. During the Second World War, Wiener developed plans for a machine to solve differential equations that looked remarkably like the electronic stored-program computers that emerged after 1945. In Britain, Turing was a regular attendee at the main cybernetics research groups, and discussed his computing ideas with leading British cyberneticians. One of those was Ross Ashby, the most prominent member of the British cybernetics community and the inventor of the homeostat machine, which used feedback to adapt to its environment, and which was dubbed the world's first artificial brain.

Cyberneticians were also interested in the idea of artificial intelligence. Wiener was convinced that machines could be designed to learn, play chess or even make decisions about military strategy. He wrote extensively about the impact that intelligent machines could have on the world of work, and about the prospect of all but the most highly skilled knowledge workers being displaced by machines. Ashby believed that the major economic and social problems of the age could be solved not by relying on a few individuals with very high intelligence, but by using computers to amplify the intelligence of ordinary people.[18] The fields of cybernetics and AI ultimately diverged, but both emerged from the shared project of creating an artificial brain.

To understand what led from the creation of 'artificial brains' in the form of computers to the birth of AI as a field, we again have to turn our attention to Alan Turing. The particularly remarkable thing about Turing was that he began to explore the idea of AI in depth even before the first of these modern computers had actually been built.

His first public pronouncements on AI came at the start of 1947, well before the construction of the Manchester Baby. The computers then being built were designed to follow instructions directly programmed by human operators, something Turing saw as being akin to the lower functions of the human brain. But in a lecture to the London Mathematical Society, he described how future stored-program machines could evolve beyond just following orders. They could, he argued, be instructed to *modify their own programs* to find ways to carry out their calculations more effectively. This was the very thing Ashby claimed the homeostat was doing around the same time.

Such a shift would effectively mean that machines were replicating the higher parts of the human brain, not just the lower. 'In such a case,' Turing argued, 'one would have to admit that the progress of the machine had not been foreseen when its original instructions were put in. It would be like a pupil who had learnt much from his master, but had added much more by his own work. When this happens I feel that one is obliged to regard the machine as showing intelligence.'[19]

Over the next few years, Turing followed up this speech with a series of articles and radio broadcasts expanding his ideas about AI. He discussed how computers could be taught to play chess; how they could learn by experience and originate new ideas; what kind of human-like functions AI would be able to carry out; what impact they might have on human society; and the objections that could be made to the idea that machines were really displaying intelligence.

This culminated in his seminal 1950 paper 'Computing Machinery and Intelligence', in which he laid out the idea of what became known as the 'Turing test'.[20] The test was based on a parlour game called the imitation game, in which an interrogator in one room had to guess which of two players in another room was a man and which was a woman, based only on their type-written responses to questions. Turing's idea was to modify the game so that one of the two players was a computer, and the challenge for the interrogator would be to tell which of them was human.

Turing saw this test as a model for answering the question 'Can machines think?' Expressed in that form, he felt, the question was so confused as to be almost meaningless. Instead, he argued, the best way to answer the question would be to replace it by another: 'Are there imaginable digital computers which would do well in the imitation game?' If the computer could successfully pass itself off as a human, then we should regard it as something that can think. There was absolutely no way that the computers that existed in 1950 could even get close to doing this. But Turing predicted that, by the end of the 20th century, they would be able to win the game. The Turing test has been used ever since as a (controversial and disputed) way to gauge whether computers are truly developing intelligence.

The origin story of AI as a discipline has traditionally been dated to 1955. It was in that year that a group of pioneering American computer scientists led by John McCarthy applied to the Rockefeller Foundation for a grant to fund an AI summer school at Dartmouth College. The phrase 'artificial intelligence' wasn't widely used at the time, and McCarthy deployed it both as a big, dramatic term to attract support and attention, and as a way to distinguish it from similar fields, such as cybernetics.[21]

In doing so, he was tapping into the prestige that the idea of intelligence had attained in Western societies. Thanks to the evolution of ideas in the 18th and 19th centuries, and the emergence of intelligence science in the early 20th century, intelligence had come to be viewed as a discrete, well-understood capability that defined the worth of human beings and drove social and scientific progress. What better way to prise money from funders than to tap into this powerful idea – the highly successful brand that intelligence had become – by promising to build a shiny new *artificial* intelligence?

But as we've seen above, the idea of AI – that the new generation of computers were artificial brains that could really be made to *think* like humans did – emerged long before McCarthy's grant application. AI wasn't something that was thought up after the invention of the computer; the two things were always bound up together, part of the same historical quest to build an artificial brain.

Fearing superintelligence

Underlying this quest to build an artificial brain was the assumption that the creation of some form of machine intelligence would be a good thing, a desirable outcome that mankind should strive towards. But the prospect of such an intelligence actually being created prompted a wave of fear about its possible consequences.

In a way these fears were nothing new. The ancient notion of a divine intelligence transcending the meagre limits of human understanding was often regarded as terrible and threatening. Many of the pre-modern stories of living machines contained an element of danger, like the golem breaking free and going on a murderous rampage. But from the late 19th century new fears began to emerge about the prospect of some kind of superintelligence.

This imagined superintelligence took different forms, including the fear of intelligent aliens, but it was most frequently associated with some variation of AI running out of control. It was a fear that was shared and stoked by many of the pioneers of computing and AI we've been discussing in this chapter. These 'expert' fears, though, were always entangled with the imagination of science fiction writers and filmmakers. It was they who first began to think seriously about what the consequences of such an intelligence might be.

The first modern treatment of the dangers of machine superintelligence came from an unlikely source. Samuel Butler was an English novelist whose major work dealt with ecclesiastical life in Victorian England. He was also deeply interested in the work of Charles Darwin and his theory of evolution. In 1863, just four years after the publication of Darwin's *On the Origin of Species*, he wrote an essay for a local newspaper in New Zealand, where he was then living. Entitled 'Darwin among the Machines', the essay raised the possibility that Darwin's account of selection and evolution in the natural world was also taking place in the world of machines. The incredibly technological progress of the steam age, Butler argued, was in effect a process of evolution in which machines were becoming self-regulating, self-acting and self-perpetuating. This was something he urged humans to take seriously. Its inevitable end, he warned, was the subjugation of the human race to the greater power of machines, just like horses and dogs had become subjugated to mankind. 'Our opinion is that war to the death should be instantly proclaimed against them,' he concluded. 'Every machine of every sort should be destroyed by the well-wisher of his species.'[22]

Butler's call to arms was somewhat tongue in cheek. But the idea of machine intelligence was obviously something he took seriously, and he returned to it a decade later in his book *Erewhon*.[23] A slightly odd, *Gulliver's Travels*-style novel, the book is an account of the land of Erewhon, which is discovered by accident by a young English traveller. Upon arrival, the traveller is instantly thrown into prison for possession of a watch. The reason, we later discover, is that Erewhon had undergone an anti-machine revolution centuries earlier. Convinced by the kind of arguments Butler had put forth in his original essay, the Erewhonians had decided that the only way to prevent their future enslavement was to destroy all machines invented in the preceding 271 years ('a period which was agreed upon by all parties after several years of wrangling as to whether a certain kind of mangle which was much in use among washerwomen should be saved or no'). The revolution had been followed by a pro-machine counterrevolution and a vicious civil war, but modern Erewhonians were convinced that their uprising had been necessary to save the human race.

Butler's novel proved hugely influential in the development of ideas about AI and its dangers. It was read by Turing, and was mentioned in some of his early writings on AI. It would later inspire the notion of the 'Butlerian Jihad' in the *Dune* novels.

It wasn't just the prospect of machine superintelligence that worried late-Victorian writers. In his 1898 novel *The War of the Worlds*, the pioneering science fiction writer H.G. Wells described the invasion of southern England by a terrifying fleet of bug-eyed, betentacled Martians wielding heat rays and poisonous gas. Wells's Martians were characterised by their huge brains, having evolved beyond the need for extraneous features like digestive systems or biological limbs, and possessing 'intellects vast and cool and unsympathetic'.

Like Samuel Butler, Wells used the language of evolution to think about non-human intelligence, and drew parallels with man's mastery of the natural world. 'We men,' the book's narrator remarks about the Martians, '. . . must be to them at least as lowly as are the monkeys and lemurs to us.'

Wells also framed his discussion about the risks of super-intelligent beings within the history of European imperialism. As we saw in Chapter 2, the modern idea of intelligence emerged alongside the expansion of European empires and the Western belief in fundamental intellectual differences between races. This kind of thinking was mocked in the novel, with the narrator remarking that, before the Martian invasion, those who thought about the possibility that there was life on Mars assumed that aliens would be less intelligent than man and 'ready to welcome a missionary enterprise'.

But Wells also offered a more scathing reflection on the way White Europeans used the idea of their intellectual superiority to justify the subjugation of others. '[Before] we judge of them [the Martians] too harshly,' Well's narrator notes, 'we must remember what ruthless and utter destruction our own species has wrought, not only upon animals . . . but upon its inferior races. The Tasmanians, in spite of their human likeness, were entirely swept out of existence in a war of extermination waged by European immigrants, in the space of fifty years. Are we such apostles of mercy as to complain if the Martians warred in the same spirit?'[24]

Following the path set by *Erewhon* and *The War of the Worlds*, hyper-intelligent little green men became a staple of science fiction in the first half of the 20th century, alongside robots, cyborgs and other intelligent creatures.

But with the invention of modern computers from the 1940s, these stories and the ideas behind them suddenly came to seem much more real. Debates about the dangers of superintelligence began to cross over from the realm of science fiction into science proper.

Ross Ashby was one of the first computer scientists to explore the risks intelligent machines might pose. Writing about the idea of a genuinely intelligent mechanical brain in 1948, he speculated on what might happen as something like his homeostat machine developed. To begin with, he argued, the machine's human trainers would be careful to ensure that it was acting in ways that benefitted them, using it to help solve the most pressing social, economic and scientific problems of the day. But as its powers developed, the machine would begin to think about how it could secure its own welfare. Would we become alarmed, Ashby asked, if it proposed burying itself deep in the ground for safety, permanently locking its power supply switches to 'on' for security? Would we do something when it began to co-ordinate its own supplies or develop its own atomic energy system? Would we notice if it realised it no longer needed humans around?[25]

These speculations were echoed by other pioneers of computing, including Alan Turing. But it was one of Turing's Bletchley Park colleagues, a bespectacled British mathematician by the name of Irving John Good, who made the next leap in thinking about what he called 'ultraintelligent' machines in an influential 1966 journal article.

Ultraintelligence was not something Good thought we should necessarily fear. Indeed, he argued that the survival of man might depend on it. But he saw the process panning out in a way that few had done before. Once an ultraintelligent machine had been built, he argued, that machine would then be in a position to design even better machines. That would lead inevitably to an 'intelligence explosion'. 'Thus the first ultraintelligent machine is the *last* invention that man need ever make,' he argued, adding the slightly worrying caveat, 'provided that the machine is docile enough to tell us how to keep it under control.'[26]

One of the most interesting aspects of Good's argument was the explicit debt he acknowledged to science fiction. Even the leading computing and AI experts, he worried, were overlooking the possibility of an intelligence explosion and its consequences. The only people who had really considered it up until then were science fiction writers. It was these writers who were raising questions about whether machines would make humans redundant, whether machines could feel pain, what social problems they would cause or solve, and how humans might form relationships with them. Serious-minded scientists might be reluctant to enter into such speculations. But, Good argued, 'it is sometimes worthwhile to take science fiction seriously'.

Good's views are a great illustration of the way that ideas about superintelligence have straddled the line between science and science fiction. Writers and filmmakers drew from cutting-edge scientific theories, and scientists in turn drew inspiration and ideas from works of fiction. AI has always been a product of fantasy and culture as much as science and technology.

Just a few years after the creation of the first computers, for example, Kurt Vonnegut drew on his knowledge of the field to think about how these new technologies might shape human society. Vonnegut would become best known for *Slaughterhouse-Five*, a book which drew on his experience as a POW during the firebombing of wartime Dresden. But his first novel, *Player Piano*, was inspired by his post-war work as a PR man at General Electric's world-leading research laboratory. His job was to explain the company's cutting-edge machines, and the ideas behind them, to the American public. He was deeply impressed by these machines, but worried about how little thought their creators were giving to their impact on human society.

Player Piano was a dystopian story about an American society of the near future in which almost all human labour had been displaced by advanced machines. As in Aldous Huxley's *Brave New World*, Vonnegut's novel integrated ideas about IQ and intelligence testing into his dystopian vision, with children's intelligence and aptitude constantly tested to determine their education and career paths. A minority in this society are earmarked for elite technical education to prepare them for influential roles as senior engineers whose job is to design and care for the machines. Everyone else is consigned to low-status manual jobs in the army or public works programmes, segregated off from the elite engineers and losing all sense of dignity and self-worth from their labour. Echoing *Erewhon*, the novel ends with a popular rebellion against the machines, although this one swiftly fizzles out when people start compulsively rebuilding the very machines they had been attacking.

Vonnegut was less interested in the existential threat of superintelligent machines than in their social and psychological impact. Explicitly drawing on the arguments of Norbert Wiener, the novel's protagonist describes the progress of the First Industrial Revolution that displaced human muscle power, and the Second Industrial Revolution that displaced routine mental work. Now that society had developed the EPICAC XIV – a reference to the real-life ENIAC computer – a Third Industrial Revolution was in the process of displacing all human thinking. This threat is personified in the novel by the character of Bud Calhoun, a gifted engineer supremely satisfied with his lot. Calhoun gives little thought to the consequences of the machine he works with, until the day he invents an ingenious new machine that makes his own work superfluous. He is promptly sacked by the automated personnel system. Little-read at the time of publication, *Player Piano* has gained renewed attention recently as a harbinger of the threat of AI automation.

As computing technologies rapidly developed in the 1950s and 1960s, more and more works of science fiction moved beyond the social dangers posed by intelligent machines to focus on their existential threat. The best-known example is the 1968 film (and accompanying novel) *2001: A Space Odyssey*. Co-written by the director Stanley

Kubrick and the science-fiction writer Arthur C. Clarke, the film drew on some of Clarke's earlier works on advanced alien life. But at the heart of the story is the character of HAL, or the HAL 9000 supercomputer, which malfunctions and kills most of the crew before it can be disabled. Kubrick consulted with Irving John Good on the HAL storyline, and the film reflected Good's ideas about intelligence explosions and the potential loss of human control. It was part of a new era of science fiction preoccupied with the dangers and possibilities of superintelligent machines, from the stories of Isaac Asimov (who developed 'Asimov's Laws' about how superintelligence could be controlled) to films such as *The Terminator*.

Such fears of superintelligent machines, it should be noted, were a particularly Western preoccupation. Western, particularly Anglo-American, stories about AI have often focused on the question of dominance and control, and on the dangers of humans losing control over intelligent machines. But these kinds of stories aren't universal. Elsewhere, science fiction and other tales of AI have drawn on different religious traditions, attitudes towards technology, and ideas about human knowledge and intelligence. The result has been a huge diversity in global AI stories, from the Russian tradition of the funny robot, to the emphasis on machine *emotions* rather than machine *intelligence* in Indian AI stories, and the idea of machines as partners to humans, rather than threats to their existence, in places like Japan and China.[27]

This diversity, however, has done little to disrupt Western fears about the threat of superintelligence. The idea that has come to be most closely associated with that threat today – the *singularity* – was also the brainchild of a science fiction writer. As in the works of Arthur C. Clarke, Vernor Vinge's science fiction was inspired and informed by his academic background in mathematics. He first coined the term 'singularity' in a science-fiction magazine in the early 1980s, expanding on it in a NASA-sponsored conference on science and cyberspace in 1993.[28]

Vinge's arguments about the singularity drew on Good's ideas about an intelligence explosion. At some point over the coming decades, Vinge argued, the world would be transformed by the creation of greater-than-human intelligence. Because this would turbocharge progress and would inevitably involve the creation of still more intelligent entities, it would almost immediately spiral beyond human control and understanding. Vinge didn't think this was only possible in the form of AI; he also thought it could develop through the emergence of superintelligence within large computer networks, or through computer/human interfaces which could substantially amplify human intelligence (a theme we'll hear more about in the next chapter). Like writers before him, he saw this change in evolutionary terms, comparable to the moment when the evolution of human intelligence fundamentally divided human beings from animals. The singularity would mean entering a 'post-human' world, the next stage of planetary evolution.

The idea of creating an artificial brain was never a straightforward one. While many of the pioneers of computing approached it as a technical question or an engineering challenge, the drive to create a machine that can think was rooted in ancient traditions of human storytelling about non-human intelligence. Once it seemed like computing technologies might turn these stories into reality, fears about dominance and control, hierarchy and evolution, quickly began to attach themselves to the new machines. The prospect of AI was bound up with humans' fears about their own intelligence, status and worth.

William Shockley and the dark origins of Silicon Valley
Let's return to the 1940s and 1950s – the period when people like Alan Turing and Ross Ashby were seeking to create the first artificial brains. It is important to remember that none of the computers that were built in that era were even remotely powerful enough to do any of the things the early pioneers of AI were dreaming up. If these machines were artificial brains, they weren't very smart ones.

The invention that transformed the room-sized behemoths of the early 1950s into the supercomputers of today, and thus helped to transform AI from idea to reality, was the transistor. And the history of the transistor reveals some of the dark currents that flowed beneath the birth of the computing industry and its spiritual home in Silicon Valley.

I knew that if I was trying to understand the history of intelligence, at some point I would have to visit Palo Alto. Once a small town built around Stanford University, wedged between pine-covered mountains and the glittering waters of San Francisco Bay, Palo Alto was transformed in the late 20th century into the sprawling heart of the computing and technology industries which became known as Silicon Valley. As such, it was at the centre of the development of the 'artificial brains' of the early computing industry, as well as later research into AI.

It was also at the heart of the 20th-century history of intelligence science. It was here that Lewis Terman worked all his academic life, at the institution that had given its name to the Stanford-Binet intelligence test which had had such a transformative impact on the early 20th-century world. And, as we saw in the first chapter, Palo Alto was central to the emergence of a new generation of ideas about cognitive elites which developed around the tech industry at the end of the 20th century and the beginning of the 21st.

So off to Palo Alto I went. On my first day I visited the campus at Stanford, wandered around Terman Park (named somewhat to my disappointment, not after Lewis Terman but his son, Frederick, who was a provost of Stanford) and headed into downtown Palo Alto. There, later that evening, I rounded a corner and was confronted by a strange, looming set of giant statues in the middle of the sidewalk. The statues looked a bit like huge spiders, or the Martian machines from H.G. Wells's *The War of the Worlds*. But on closer inspection, they turned out to be giant models of electronic components – diodes and transistors.

As the plaque on a nearby wall explained, the sculptures were monuments to the legacy of Shockley Semiconductor

Semiconductor sculpture. Birthplace of Silicon Valley, 2018 by Mary Bayard White and Vickie Jo Sowell. Photo: Dicklyon. CC BY-SA 4.0

Laboratory, which had been founded on that site at 391 San Antonio Road in 1956. This, the plaque explained, made it the 'birthplace of Silicon Valley'. A diagram underneath depicted a kind of Silicon Valley family tree, showing how the people and ideas developed at the Shockley Semiconductor Laboratory had gradually spread out, inspiring and peopling the famous tech companies that had made Palo Alto their home over subsequent decades.

Shockley Semiconductor Laboratory had been founded by William Shockley. In the late 1940s Shockley had helped to invent the transistor, the foundation of modern electronics. For this monumental achievement he was awarded the Nobel Prize for Physics. In 1999, *Time* magazine named him as one of its 100 most important individuals of the 20th century. Despite these achievements, Shockley would ultimately be remembered for the obsession with genetics, eugenics and race that dominated the final decades of his life. One of the greatest scientific figures of the 20th century had transformed himself into one of its most notorious champions of scientific racism. And at the heart of that transformation lay a life-long obsession with intelligence.

Shockley was born in London in 1910, but grew up in Palo Alto. His parents were both academically successful. His father was an academic polymath turned engineer, and his mother was one of the first women to graduate from Stanford before becoming a leading mining surveyor. From a very early age they were interested in their son's cognitive development, not necessarily believing him to be a genius but obviously proud of his progress. 'His intelligence is developing quite rapidly,' his father noted with approval in his diary when his toddler learned to count to four.[29] As a teenager in Palo Alto his mother got his IQ tested as part of Lewis Terman's gifted child project. Disappointingly, his score of 129 fell just short of the 135 needed to be accepted onto the programme.[30]

Shockley followed in his parents' academic footsteps, studying at Caltech and MIT. During the Second World War he worked for the American military, carrying out statistical work on anti-submarine warfare, aerial bombing and nuclear weapons. At the end of the war he found himself in a senior role at Bell Laboratories, the research arm of the telecommunications monopoly, AT&T.

It was here in 1947 that two members of Shockley's lab team, John Bardeen and Walter Brattain, successfully tested the first transistor. Shockley wasn't personally involved in the test, and there was some controversy over whether he should receive equal credit for it alongside his two colleagues. But he undoubtedly played a key role in directing the research projects which the transistor emerged from. And immediately after that first experiment he made some major breakthroughs of his own in the development of a different form of junction transistor. For these achievements he was awarded the Nobel Prize in 1956, alongside Bardeen and Brattain.

Transistors transformed electronics and launched the modern computer age. The earliest computers, built by the likes of Turing and von Neumann, had relied on thousands of fragile vacuum tubes acting as switches to direct the flow of electric current. Shockley's solid-state transistors did the same job more quickly and reliably, taking up much less space. Over time they would get smaller, becoming integrated into circuits on single microchips, and eventually into today's microprocessors. The first Intel chips of the 1970s contained a few thousand transistors each. By 2023, the number that could fit onto a single microprocessor had risen to 5.3 trillion. All our modern digital technologies are built on the back of Shockley's humble transistor.

After the invention of the transistor, Shockley left Bell Labs to set up his own company. His aim was to become the first mass manufacturer of silicone transistors. His decision to base the Shockley Semiconductor Laboratory just down the road from his childhood home in Palo Alto would have seismic consequences for the future of the tech industry.

The company itself failed to live up to the hopes and promises of its founder and didn't survive for long.

William Shockley celebrates the news of his 1956 Nobel Prize in Physics with his employees. Source: Computer History Museum.

Shockley hired a raft of young, cutting-edge scientists, including Gordon Moore and Robert Noyce, who would go on to find global fame as the founders of Intel. But his business strategy was flawed, and his management style left much to be desired. Eventually these management failings, including his insistence on putting everyone through a lie detector test following a minor workplace accident, proved too much for his team. In September 1957 they quit en masse to found their own company, Fairchild Semiconductor.[31]

Fairchild succeeded where Shockley had failed, and when it too dissolved in the late 1960s, its alumni went on to found many of the companies that drove the growth of Silicon Valley into the economic behemoth it is today. Among those companies was Noyce and Moore's Intel, which was the first to figure out how to integrate the entire circuitry of a computer, including its transistors, onto a single chip.

Gordon Moore is now best known for 'Moore's law', which correctly predicted that the number of transistors on a microchip would double every two years. He had witnessed Shockley's irascible management style firsthand as one of his earliest employees. But he identified Shockley Semiconductors as the start of the lineage of semiconductor companies which sparked the epoch-defining industrial boom of the late 20th century. 'It was Shockley,' he argued, 'who brought the silicon to Silicon Valley.'[32] It's this foundational role that is commemorated today with the giant transistor statues on a Palo Alto pavement.

Many aspects of Shockley's personality eerily foreshadowed the Silicon Valley tech tycoons of the early 21st century. For one thing, he was ahead of his time in his obsession with physical fitness. As a young man he had been heavily into workouts, calisthenics and sunlamps, and had even modelled for an exercise equipment ad. At Bell Labs he used to do pull-ups around the office and would sometimes enter a room via a back flip from the door frame. He lived before the current tech industry obsession with cryogenics and anti-ageing science. But in later life he did donate to a 'superbaby' sperm bank that only accepted donations from Nobel Prize winners, which suggests at least a passing interest in the idea of genetic immortality.[33]

He also placed great faith in the idea of measuring people's ability and productivity. This interest began at Bell Labs, where he began to think about how scientific creativity could be quantified to help ensure the recruitment of what he called 'topnotch men'.[34] At Shockley Semiconductors, potential recruits were faced with a barrage of tests, from both Shockley himself and the psychometric testing agencies he employed. Shockley clearly wanted to hire the most intelligent people in the field, but he also seemed deeply anxious lest any of them should turn out to be more intelligent than himself. He saw scientific creativity as fundamentally a team process, but one which needed to be hierarchically organised. Creativity would not emerge from below, he believed, but must trickle down from the scientific genius and their highly intelligent collaborators at the top.[35]

This interest in intelligence, and some of its more disturbing implications, was also evident in Shockley's personal life. Like his management techniques, Shockley's parenting style could be psychologically abusive. He struggled to relate to any of his children, whom he clearly regarded as his intellectual inferiors, and had almost no contact with them later in life. In a notorious interview with *Playboy* magazine in 1980 he was asked about his children in the context of his beliefs about genetics and eugenics. 'In terms of my own capacities,' he answered, 'my children represent a very significant regression. My first wife – their mother – had not as high an academic-achievement standing as I had.'[36] Shockley's obsession with intelligence underpinned the embrace of scientific racism that marred the final decades of his life, and ultimately overshadowed his prior achievements.

No one is quite sure why, in the mid-1960s, Shockley effectively abandoned his work on electronics and physics in favour of the study of intelligence, genetics and race. There is evidence that he had long held views about White intellectual superiority, but they weren't something he appears to have shared widely or thought much about. Some have speculated that the change was linked to his increasing paranoia, his alienation from his children, his failing business or a major car accident he'd been involved in in 1961. But whatever the trigger, around 1963 Shockley decided to start airing his theories about the evolutionary deterioration of the human race.

He claimed that his concerns had begun with his experiences in Bengal, while working with the US Army during the Second World War. The poverty and overcrowding there, he said, prompted him to start reading about the risks of global overpopulation and food supply. In this he certainly wasn't alone. The middle of the 20th century witnessed an explosion in the size of the global population, and there were widespread concerns at the time that this would inevitably lead to mass famine and economic underdevelopment.

But for Shockley, these concerns merged with what he saw as disturbing trends closer to home. The story he returned to again and again in his talks and writings was an attack on a store owner that took place in San Francisco. The store owner had been blinded by an acid attack. The acid-thrower, who had been hired by someone with an unjustified resentment against the store owner, was a teenager who was, in Shockley's words, 'one of approximately a dozen illegitimate children of an irresponsible and destitute woman'.[37]

The conclusion Shockley drew from this incident was that the affluence of modern American society had made it much easier for women like this to have lots of children. 'This brought home to me,' he told an audience of Nobel Prize-winning scientists in 1965, 'that if we had a situation in which an irresponsible individual could produce offspring at a rate which might be four times greater than those of more responsible members of society, this was a form of evolution in reverse.'[38]

Right from the start, Shockley's embrace of eugenics sparked controversy. But his proposed solutions to these supposed problems – legalisation of abortion, promotion of international birth control policies, better scientific education – weren't at this stage that far beyond the mainstream. That quickly began to change when Shockley turned his attention to the relationship between race and IQ.

In an interview with *U.S. News & World Report* in November 1965, Shockley claimed that the average Black IQ was lower than the average White IQ. He suggested that this explained the high crime and welfare rates among African Americans. At this stage these ideas formed a relatively minor part of Shockley's overall argument about genetic decline. But they gradually began to feature more and more. The story of the store owner and the acid attack was elaborated to reveal that the assailant was Black and the victim was White. Much more detailed arguments began to emerge in his writing about racial disparities in IQ, and their causes and consequences. And Shockley launched a full-throated campaign against the American scientific establishment, which, he claimed, was too scared to investigate these subjects properly.

Shockley was launching these arguments during the peak of the Civil Rights movement, the emergence of Black Power, and the campus revolts against the Vietnam War. Unsurprisingly, they provoked a huge backlash. Shockley seemed to relish the attacks he faced, arranging debates with his critics, publicising attempts to prevent him from speaking on campuses and challenging criticisms in the press. As he became more and more embroiled in the culture wars, his obsession with race and IQ became ever more all-consuming, and his arguments more extreme. By the early 1970s he was arguing that each 1 per cent of White blood increased an individual's IQ by one point.[39]

Shockley's arguments about race and eugenics were built around a core set of beliefs concerning human intelligence.

He believed that the unprecedented economic and technological progress of the past two centuries had been driven by the application of human intelligence to the shared problems of mankind. That process, he argued, had reached its peak in the modern United States. The importance of intelligence to modern society was reflected in the fact that intelligent people tended to do better in life.

But Shockley felt that mankind had now reached a tipping point. The great progress that human intelligence had achieved had created enormous new challenges, like the threat of nuclear annihilation. The only way to forestall catastrophe was to successfully apply human intelligence to the task of solving these challenges. But modern affluence and the welfare state was undermining these efforts. The 'least fit' in society were no longer being eliminated as they once were, and the genetic decline which resulted threatened mankind's intellectual resources. This was particularly a problem in the United States because, he claimed, the decline in intelligence was more pronounced among Black Americans. The only way to address this decline was to apply human intelligence and reason to the questions of reproduction and genetics – in other words, to consider mankind as a biological species like any other, and approach its breeding in the same way one might the breeding of horses or dogs.[40]

William Shockley died in 1989, largely unknown and unmourned, estranged from his children, and with his scientific legacy indelibly tarnished by his embrace of scientific racism. His extreme views on race and eugenics weren't shared by most of those working in the new fields of computing or AI. But his foundational beliefs about intelligence and its importance weren't a million miles from the mainstream. The modern computing industry that Shockley helped to launch, the digital revolution that followed and the emergence of AI were all bound up with ideas about what intelligence was, and why it mattered.

Computers were created as artificial brains, designed to carry out certain mental tasks more quickly than human brains could, with the hope that they would eventually be able to think and learn just like we do. Those who pursued this goal did so on the assumption that human intelligence could be understood and replicated, and that it was something that society needed more of to meet the challenges of the era.

Right from the start there were those who saw computers and AI as something positive, amplifying human intelligence and handing us the analytical tools needed to solve the most complex questions posed by the modern world. Others saw the rise of intelligent machines as a threat, worrying about a future of mass unemployment or the existential risks that AI could pose to mankind. What both perspectives shared was the idea not only that intelligence was a key part of modern life, but that it was going to become even more important in the future.

8

Augmenting Intelligence

'... each one of us is potentially Mind at Large'.
Aldous Huxley

IN 1982 A YOUNG Venezuelan army lieutenant named Hugo Chávez made his first ever appearance on national television. He was being interviewed as one of the graduates of a new programme designed to improve the intelligence of army officers. As part of the programme, Chávez and his fellow officers had studied a series of thinking skills developed by Edward de Bono, the world-renowned business writer most famous for his concept of 'lateral thinking' and bestselling books like *Six Thinking Hats*. The programme had been created by Venezuela's Ministry for the Development of Intelligence, a short-lived but pioneering experiment in augmenting human intelligence. The ministry had been set up under a centre-right government and was led by various members of the Venezuelan upper bourgeoisie. Less than two decades later, Chávez – by now an experienced and charismatic television performer – was elected as president, overthrowing the bourgeois state and ushering in the Bolivarian revolution. Sadly, he never revealed whether de Bono's lateral thinking skills had facilitated his rise to power.[1]

The idea behind the Venezuelan project was that intelligence could be *augmented*. This was certainly not a new idea. The entire point of the education system, after all, is to develop people's knowledge, skills and mental abilities, to augment the power of the mind beyond what it would be capable of in its uneducated form. And as we've seen in previous chapters, early 20th-century eugenicists believed that mankind's intellectual qualities could be improved through policies to control human reproduction.

But the idea that intelligence *could* and *should* be augmented boomed in the second half of the 20th century. During the first half of the century, the modern idea of intelligence and its importance had spread around the world, and embedded itself in everyday life and popular consciousness. The consensus at this stage was that intelligence was largely inherited, rooted in people's biology and genetics. But the middle of the 20th century witnessed a turn away from this idea. Although the issue remained contested and controversial, the debate around intelligence shifted to emphasising the environmental forces that shaped people's intelligence levels. The quality of a person's intelligence came to be seen less as something rigidly fixed at the moment of their birth, and more of something that could develop and grow over time with the right support.

For those who subscribed to these ideas, the logical next step was to think about *how* exactly people's intelligence could be encouraged to grow and flourish. It was this line of thinking that led to the late 20th-century boom in efforts to augment human intelligence. A host of new ideas and projects emerged, aiming to augment human intelligence through education and parenting, thinking skills and techniques, or new technologies and drugs. This is the history we're going to explore in this chapter. It's a story of Latin American army officers, naked hippies, hot-housed children, global business gurus, cyberneticians, computer geeks, drug users, neuroscientists and megalomaniacal billionaires. And it's a story, as we'll see, that runs right up to the present day.

Human potential

The development of psychology in the 19th century was accompanied by new ideas about the human mind and human intelligence. While some people argued that intelligence was fixed and that human abilities were tightly bound by heredity, others believed that this new science of the mind opened up new possibilities for individuals to enhance their mental powers. By thinking more effectively, they argued, human intelligence, or at least its effectiveness, could be radically improved.

These ideas first emerged in literature on 'mental hygiene' in the late 19th and early 20th centuries. The field of mental hygiene generally focused on psychiatric and mental illnesses, advocating for better care and treatment of the mentally ill. But it also engaged with a broader set of ideas about mental well-being. Improving mental hygiene, some began to argue, could free human minds to work more effectively and thereby boost individuals' cognitive power.

As the new science of intelligence testing spread after the First World War, writers and publishers began to realise there was serious money to be made in books promising to teach people how to boost their intelligence. By the 1930s a vibrant new genre of mental 'self-help' literature had emerged, a genre that was particularly popular in the United States. Books with titles like *Use Your Mind: The Road to Successful Thinking* introduced readers to some of the new theories of mind emerging from psychology, and offered techniques to improve understanding, comprehension, judgement and memory.[2] The economic boom of the 1920s and the Great Depression that followed fuelled demand for books that told people how to use the powers of their mind to achieve personal and professional success – in the words of one bestseller from the time, how to *Think and Grow Rich*.[3]

From the 1940s publishers began to move into books aimed explicitly at business audiences. The American advertising executive Alex Osborn began to publish books in the late 1930s promising to show how teams could work more effectively together to improve their creativity. It was he who introduced the world to the idea of 'brainstorming', which promised to counter negative thinking and solicit new ways to solve problems.[4] By the 1950s those ideas had taken the American business world, so to speak, by storm.

As the 1960s loomed into view, another set of ideas about maximising the powers of the human mind began to emerge, in the form of the Human Potential Movement. This movement owed much to the writings of Aldous Huxley. Huxley was a British author who had become famous in the 1930s for his dystopian novel *Brave New World*. Just before the Second World War he had moved to southern California, where he lived until his death in 1963. During the final decades of his life he helped to foster the emergence of the counter-culture movement, studying various forms of Eastern spirituality and writing about his experiences with the psychedelic drug mescaline in his hugely influential book *The Doors of Perception*.

In 1960 and 1961, Huxley visited a number of US universities to explain his ideas about human potential. Humankind, he believed, was on the verge of unlocking abilities which had so far remained latent. This wasn't a call for eugenics, something he'd written about in *Brave New World* and which had been enthusiastically promoted by his brother, the biologist Julian Huxley, as the first head of UNESCO. Instead, he believed that human beings *in their current genetic state* had the potential to unlock vastly greater powers of reason, emotion, creativity and kindness. Neuroscientists, he argued, had shown that humans only ever used 10 per cent of the neurons in their brains.* Whether through new forms of education, therapies or drugs (we'll return to them later), it was now becoming possible to unleash the power of the other 90 per cent.[5]

Among the audience at one of Huxley's talks were the founders of Esalen, the community in Big Sur, California, which became the spiritual home of the human potential movement. The goal of Esalen was to help people achieve the transformation of human potential which Huxley had spoken about. From the early 1960s, tens of thousands of people visited the centre, seeking emotional growth, self-actualisation and transcendence. A visit to Esalen usually involved so-called 'encounter groups' – marathon discussion sessions in which people were encouraged to share details of their lives and receive (often pretty brutal) feedback from their peers – alongside group massages, yoga, communal bathing and not a little nudity.

Esalen was not, however, just a hippie holiday camp. It was led by some of the most high-profile humanistic

---

* This is an idea you may well have heard of before. It's not clear where it first came from, but the human potential movement played a big role in popularising it in the 1960s. It was a key plot point in the 2014 movie *Lucy*, in which Scarlett Johansson's eponymous heroine is exposed to a new drug which allows her to access 100 per cent of her brainpower, in the process gaining superhuman strength, communicating telepathically, controlling matter, morphing into a supercomputer, and eventually disappearing into some kind of spacetime continuum. Sadly, the idea that we only use 10 per cent of our brain is a myth.

psychologists in the country, including Abraham Maslow, who was the president of the American Psychological Association and creator of Maslow's hierarchy of needs. Esalen, Maslow argued, was 'potentially the most important educational institution in the world, a place where modern psychology was used, not just to heal the mentally sick, but to improve the already well'.[6] This idea of improving the well was a central plank of the human potential movement.

Esalen was not the kind of environment where people commonly talked about their IQ or boasted about their brainpower. Its focus tended to be on emotional rather than intellectual development. Some critics even charged it with being anti-intellectual, prioritising the gut over the brain. But others in the United States took the ideas of 'improving the well' and maximising human potential and tried to apply them much more directly to the concept of intelligence.

Chief among them was the psychologist Glenn Doman. Doman had begun his career working with brain-damaged children, developing a popular but controversial technique which he claimed could help his patients regain brain function. In the 1940s, he established the Institutes for the Achievement of Human Potential in Philadelphia, which developed this work on brain injuries. But like the psychologists at Esalen, Doman believed that his techniques could benefit the well just as much as the sick. In particular, he believed they could be used to supercharge the intelligence of very young children.

Doman described the supercharging of infant intelligence as a 'gentle revolution'. His mission, he declared, was to 'give all parents the knowledge required to make their babies highly intelligent, extremely capable, and delightful children, and by doing so to make a highly humane, sane, and decent world'.[7] Like all the other people who began to publish intelligence self-help books in the middle of the 20th century, he realised there was money to be made from promising people the secrets to greater brainpower.

For a not-inconsiderable fee, Doman invited new and prospective parents to attend seven-day workshops at his Better Baby Institute. There they would learn the science and secrets of how to make their babies superintelligent. Doman and his team told parents that every child, at the instant of birth, had a higher potential intelligence than Leonardo da Vinci. The first six years of life, he argued, were crucial for unlocking this potential. If parents followed his programme, their children could begin reading at 8 months, playing chess at 18 months and writing stories at 28 months, as well as mastering music, maths, sports and general knowledge.

Doman had huge faith in high intelligence and its power to change the world. Like the intelligence scientists and gifted education pioneers who had gone before him, he fought against the idea that intellectual geniuses were unstable misfits. Indeed, he told parents that the most intelligent children were also more self-sufficient, happier, less violent, more active, kinder and gentler. The more of them there were in the world, the better the world would be.

The flip side of this argument was that, according to Doman, 'it is the least competent, incapable, dull, insensitive, unknowing kid who whines, cries, complains, and hits'.[8] Woe betide those in Doman's world who either weren't intelligent themselves, or who refused to do everything in their power to make sure their children were. Indeed, Doman was particularly critical of nurseries and kindergartens, and of the parents who chose to send their children to them. He believed that a baby's intelligence could only be nurtured through the direct and loving attention of parents. Not incidentally, his books almost always referred to 'mothers' when he discussed the importance of this attentive, face-to-face parenting.

Doman denied, however, that he was an elitist. His ambition was to turn all of the 1 billion children then alive on the planet into a superintelligent elite, superior to no one but their undeveloped selves. To help spread his mission he began in the 1960s to publish a series of best-selling 'how to' guides for parents, books explaining how to teach babies to read, how to teach them maths and how to multiply their intelligence. The books have remained hugely popular for decades, selling millions of copies in dozens of languages worldwide. The behemoth that is the 21st-century parenting advice industry is still suffused with the idea that there are scientifically grounded techniques that will help boost a baby's intelligence.

It's no coincidence that the 1960s witnessed such a boom in ideas about expanding human minds and transforming human intelligence. The early intelligence scientists at the start of the 20th century had emphasised the genetic nature of intelligence, arguing that it was mainly an inherited characteristic. But the period from the 1950s to the 1970s saw a turn away from hereditarianism towards the environment – from nature to nurture – with psychologists increasingly emphasising the ways in which intelligence was shaped by a person's childhood, family, education, social background and so on. The logical corollary of this was that intelligence, rather than being fixed at birth, could be improved through changes to a person's environment. Combined with the (partial) turn towards egalitarianism in Western politics after the Second World War and the new language of human potential flourishing in the 1960s, this scientific turn towards environmentalism encouraged the popular idea that people's intelligence could be improved.

Thinking skills

When Hugo Chávez reminisced in later life about his experiences with the Venezuelan Ministry for the Development of Intelligence, he recalled being taught techniques with names like PMI (Plus, Minus, Interesting) and CAF (Consider All Factors).[9] These tools were meant to boost the decision-making powers of Venezuela's junior officers, helping them harness the power of their brains to process information and make better decisions. Such techniques would have looked very familiar to a global business executive of the 1980s and 1990s. They had been developed by a British-Maltese business guru called Edward de Bono, the man who did more than anyone else to promote the idea of 'thinking skills' in the second half of the 20th century.

De Bono was born into an elite Maltese family just before the Second World War, growing up in a household that showed little emotion or affection but had huge respect for intellectual achievements.[10] He was something of a child prodigy, entering the University of Malta at just 15 years of age and taking only five years to qualify as a doctor. He then won a Rhodes Scholarship to study at Oxford, before going on to work as a researcher at the University of Cambridge.

Despite these academic achievements, de Bono had little faith in the idea of intelligence. The problem with intelligent people, he felt, was that they knew they could defend their point of view effectively, whether it was right or wrong. Too often, therefore, they fell into the trap of just criticising others rather than being creative or constructive in their own right. The thing that he came to value above all else was not intelligence, but *thinking*. This was something he believed everyone could learn to do well. Instead of individuals thinking of themselves as intelligent or unintelligent, as someone who could or couldn't pass exams, he wanted *everyone* to see themselves as a thinker.

In the 1960s he began to transform these ideas into a series of books setting out his ideas about what effective thinking was and how it could be developed. The idea he would become most associated with was introduced in his second book, in 1967 – lateral thinking.[11] Contrasting it with rational, 'vertical' thinking, which he felt had dominated Western thought since the time of Aristotle, de Bono argued that it was lateral thinking that was the true source of creativity, the thing that allows us to generate new ideas and solutions.

Over the following two decades, de Bono would develop this idea into a series of elaborate programmes designed to guide people towards this kind of creative, lateral approach to problems. These included the famous *Six Thinking Hats* model, which became one of the biggest-selling business books of the 1980s. His ideas were hugely popular in the business world. Despite being a fairly low-key speaker (de Bono's public speaking *modus operandi* involved sitting on stage and doodling continuously on an overhead projector), he was almost constantly flying around the world to address audiences of managers and business leaders, advising some of the world's biggest organisations, from Nestlé and Ford to BP and Siemens.

He was also feted by governments, and often found himself in demand in conflict zones. He worked extensively in Northern Ireland during the 1970s and 1980s, including holding meetings with the leadership of Sinn Féin at the peak of the Troubles. He offered advice to world leaders on negotiations between Israelis and Palestinians. In the 1990s he even tried to establish a new international organisation – something he referred to as an 'intellectual Red Cross' – which would offer neutral problem-solving mediation in conflict zones around the world.

But his biggest impact was probably in the world of education. Despite his academic background he was sceptical about the value of modern education systems, arguing that knowledge-based courses were becoming obsolete, and that schools and universities should focus on teaching practical problem-solving skills which could be applied to real life. In the early 1970s he developed a programme called CoRT (Cognitive Research Trust), which was designed to embed such real-life, practical thinking skills into the educational system. The programme introduced simple tools like PMI (Plus, Minus, Interesting), which involved listing the good, bad and interesting features of an idea, or CAF (Consider All Factors), which encouraged a full evaluation of all aspects of a question before trying to decide on a solution. For de Bono, these techniques provide a scaffolding that anyone could use to think their way more effectively through a problem or situation.

Elements of his programme were introduced into education systems in Australia, New Zealand, Canada, the United States, Malaysia, Nigeria, Bulgaria and Singapore, among others. But it was Venezuela which saw the most ambitious attempt to implement de Bono's ideas.

In 1979, Venezuela decided to make it a national mission to improve the intelligence levels of the entire country. It became the only country in the world ever to create a ministry dedicated solely to the development of intelligence.

The project was inspired by a man named Luis Alberto Machado. Machado was a conservative lawyer who, in the 1970s, had become increasingly interested in the role of intelligence in modern society. Intelligence, he believed, was malleable – something everyone had the capacity to train and perfect. And he felt that, as the world became more complicated and scientific knowledge continued to develop, intellectual ability was becoming more vital to collective progress. In his first book, *The Revolution of Intelligence*, he argued that the countries that would thrive in the future were those that did the most to develop their people's intellectual capacity.[12] In a follow-up book called *The Right to be Intelligent*, he emphasised that this applied most urgently to the most marginalised groups in society, those who were often denied access to decent education and living conditions.[13] Without harnessing the creativity and intellectual potential of every single member of society, he warned, true development would never be possible.

These arguments caught the attention of Venezuela's newly elected president, Luis Herrera Campins, who in 1979 appointed Machado to head up a new Ministry for the Development of Intelligence. Machado's new ministry pursued three lines of work.[14] The first was the Family Project, which focused on infancy and preschool years. Its goal was to ensure that all children received the kind of stimulation needed to fully develop their mental capacities, alongside sufficient nutrition and healthcare. Radio and television programmes were developed to teach expectant mothers how to engage and stimulate their newborns. Trainers were placed in hospital maternity units and post-natal health clinics to share techniques. And preschool teachers were given training in how to provide a stimulating environment for children.

The second strand of the project was aimed at schools. It was built around a series of problems designed to teach children different thinking skills, and to help ensure that those skills became habitual parts of their learning and daily lives. Many of them were built around the thinking skills developed by Edward de Bono. Machado had read de Bono's work, and was introduced to him by the British media oligarch (and later disgraced fraudster) Robert Maxwell. Their ideas were not exactly aligned – whereas de Bono argued that intelligence was of limited importance and that thinking skills could enable everyone to *operate* their native faculties more effectively, Machado believed more straightforwardly that the right education could *increase* intelligence levels. But there was sufficient overlap for Machado to embed the CoRT programme into his projects. De Bono himself travelled to Venezuela on numerous occasions, delivering training to hundreds of facilitators who then went on to roll out the programme across the country.

Other strands were developed by researchers at Harvard, including by the psychologist Richard Herrnstein, who gained public notoriety for his views on race and intelligence. They were aimed at promoting reasoning, inventive thinking, problem-solving and decision-making. Similar programmes aimed to boost children's IQ through the systematic study of chess, and to improve their creativity through teaching classical music.

The final elements of the ministry's work were aimed at adults. These were also largely based around the thinking skills of de Bono and others. Programmes were developed to reach people of all ages and across all professional and social groups. They included evening classes for resident of Caracas' poorest *barrios*, teaching skills that could be used to address everyday problems; programmes for peasants aimed at increasing agricultural productivity; schemes for workers aimed at improving job and personal effectiveness; and those for civil servants and army officers designed to improve creativity and decision-making. It was the latter programme that included Hugo Chávez as an early participant. The television show he appeared on was part of a series of national broadcasts that promoted the work of the new ministry to the Venezuelan public, and introduced them to some of the de Bono's core thinking skills.

By the early 1980s, the ministry was reporting that these initiatives had reached 1.5 million ordinary Venezuelans.[15] They had become central to the government's domestic policy, written into its national development plan. Its proponents hoped that it would boost the country's long-term economic development and help create a stronger, more cohesive democracy.

But the government also had one eye on an international audience. It wanted to present Machado's project as a model for social development worldwide. The ministry received support from global personalities like the behavioural psychologist B.F. Skinner and the American polymath Buckminster Fuller, as well as institutions like the Club of Rome and Harvard University. It helped to found the Latin American Centre for the Development of Intelligence, based in Uruguay, which sought to spread its ideas across the continent. It collaborated with international organisations like the International Labour Organization and UNICEF. It organised international conferences which attracted governments, NGOs and experts from around the world. And it received interest and support from countries including Ecuador, Japan and Senegal.[16]

These grand global plans never really came to fruition. Although the ministry's programme attracted a lot of international interest, there was also scepticism about its approach. Critics questioned the theory that collective intelligence could really be increased, the real-world impact of the ministry's policies, and the assumption that better thinking skills would unlock economic and social development. UNESCO, for example, had been central to Venezuela's international lobbying efforts. Its director-general praised the idea that each person was capable of achieving a higher level of intelligence as 'profoundly humanist'.[17] But there were many influential people in UNESCO who dismissed the project as a sham, and ultimately the organisation decided to maintain a prudent distance.

They were probably right to do so. When a new centre-left government took power in Venezuela in 1984, the ministry was eliminated. After a brief period of assessment, almost all of its projects were shut down. The idea that Venezuela's development would be driven by improving the intelligence of its people was abandoned. It would prove to be the last, and indeed the only, global attempt to explicitly place the mission of increasing national intelligence at the heart of government policy.

Despite this failure, the influence of de Bono's thinking skills lived on. Many of the Venezuelans who had worked with de Bono's techniques continued to promote the teaching of thinking skills in schools into the 1990s. Some were even involved in re-exporting those skills to new educational projects in China and South Africa.[18] Across the world, including in the United States, thinking skills became an increasingly prominent part of school curriculums from the mid-1980s. Thinking skills of various sorts continue to be taught in universities, business schools and corporations around the world to this day.

Technology

Do you feel like your smartphone makes you smarter? When you sit down at your computer, does it seem like it's multiplying your brainpower? When you're drowning under an avalanche of emails, instant messages or online meetings, are you grateful for the way technology has turned you into a superintelligent problem-solver?

No? How strange! Because that's what the people who invented these technologies thought would happen. The new era of computing and digital technologies, they believed, would magnify the power of our minds, augment our intellects and help launch us into a new era of human superintelligence.

Take, for example, Ross Ashby, the man we encountered in the previous chapter who created the homeostat machine. Writing in the leading computing, AI and cybernetics journal in 1956, he set out a blueprint for what he called 'intelligence-amplifier' machines.[19]

Like many of the people exploring these ideas, he started with the problem of complexity. The modern world, he believed, was becoming more and more complex. The problems the world faced – from finding ways to feed a rapidly growing population to maintaining global peace – were now stretching the intellectual capacity of humans to solve them. It was no good waiting for some superintelligent genius to come along and solve these problems for us. The research of intelligence scientists like Lewis Terman had proved, Ashby believed, that the range of human intelligence was just too limited. Mankind in the modern era was in the same position as engineers in ancient Rome or the Middle Ages who encountered a problem requiring physical strength far greater than the strongest person. Ultimately these problems couldn't be solved until the invention of steam power, when new machines were developed to amplify the physical power of a single worker many times over. Now mankind needed to develop new machines that would do the same thing for brainpower.

Although he was happy to draw on the work of intelligence scientists and the idea of IQ, Ashby wasn't interested in explicitly defining what intelligence was. All he was really interested in was the ability to solve problems. He believed that this was ultimately a matter of selection

– the ability to choose the right solution to a problem from all of the possible solutions available. And this, he believed, was something that the new generation of electronic thinking machines would soon be able to do. Devices, supplied with a small amount of intelligence from a human operator, would be able to emit a much larger amount. They would amplify intelligence in the same way that a music speaker amplifies sound. The effect would be like having a human with an IQ of a million!

Ashby was interested in the nitty-gritty of how these machines might work on a theoretical level. He was, after all, still working in the computers-the-size-of-a-room era, and there seemed to be little immediate prospect of his vision becoming reality. But other people were thinking along similar lines, and were exploring how this new generation of machines might work in the real world. How would human and machine actually have to interact for this intelligence amplification magic to happen?

One of the first people to really grapple with this problem was the American scientist Vannevar Bush. Bush had a background in electrical engineering, and had been a pioneer of analogue, electromechanical computing in the in 1920s. During the Second World War he had been put in charge of co-ordinating the 6,000 scientists who were drafted to work on military projects for the US government, including the development of the atomic bomb.

Like Ashby, Bush was concerned about complexity. His experience as a science administrator had made him acutely aware of the sheer scale of modern scientific and technical knowledge. The age of the 19th-century gentleman scientist who could keep abreast of the latest developments across different fields was long gone. Now the sheer scale of research being produced by scientists around the world was far beyond what a single person could hope to read. This was leading to increased specialisation, with individual scientists only able to keep up with the cutting edge of narrow sub-disciplines. This, Bush feared, created a huge risk that vital advances in one discipline would be overlooked in other disciplines where they could be applied. Added to this was the fact that new technologies, particularly computers, promised to expand the amount of knowledge and research at an even greater rate in the future. Humanity was producing mountains of new information, but all that information would only do any good if it could be found, identified and utilised by the people who could understand it.

As the end of the war approached in the summer of 1945, Bush decided to bring his concerns to the attention of the general public, and particularly to the scientific community, with a famous article in the *Atlantic* magazine entitled 'As We May Think'.[20] Speaking to the thousands of scientists who were being released from government service back to their own research, he urged them to turn their attention to the question of information and how it should be used.

Bush's article painted an eerily prescient picture of the way scientists and other knowledge workers would come to use technology. He imagined a scientist of the future moving around their lab, taking hundreds of pictures with a miniature camera attached to their forehead and operated by a shutter in their hand. This scientist would speak their comments out loud. Their words would be automatically recorded and time-linked to their photos, and miniature copies would be stored for future reference. Computers, fax machines, televisions and microfilm would help them to carry out their work more easily and efficiently. Like Ashby, Bush saw this technology as a means of enhancing intelligence. As more of the repetitive and logical work of scientists could be automated, so their brains would be free to explore their intuitions and apply their creativity.

But Bush's vision went beyond the scientist working in the lab. His article came to be remembered for introducing the idea of what he called the 'Memex' machine. The Memex was a desk, with a slanting screen on top surrounded by a keyboard and various buttons and levers. At one end of the desk there would be a storage facility for microfilm copies of books, newspapers, letters and so on, with capacity for millions of pages. The user sitting at the desk could call up any document they wished on their screen, read it, amend it or take notes, and save an image of those notes to storage. Calling up information could be done via a basic search function, but there would also be a system of 'associative indexing', whereby users could link any document to any other, with those links then recorded and saved so the same 'trail' could automatically be followed in the future, either by the same user or by any other user they shared it with.

The Memex machine was never built. But at a time before the first electronic digital computers had been developed, Bush was imagining a world of personal computing, digitisation, hypertext and multimedia. Indeed, the computer scientist who coined the term 'hypertext' in the 1980s, Ted Nelson, was explicit about his debt to the vision of the Memex.[21]

Bush saw all of this as a process of intellectual enhancement. Like so many other computing pioneers, he drew parallels between his Memex machine and the human brain. He saw 'associative indexing' as a model for how the human mind connected ideas and information. He saw the microfilm storage system as an improvement over unreliable human memory, granting people the 'privilege of forgetting', and freeing up their minds to explore new ideas. And he believed ultimately that computers and machines like the Memex would enable mankind to improve its thinking processes, magnifying the power of the human intellect and driving civilisation forward.[22]

A few months after Bush's article was published, a copy found its way into the hands of a young US Navy radar technician in a Red Cross library on the edge of a jungle in the Philippines.[23] Inspired by Bush's vision, Douglas Engelbart would go on to build a career trying to turn the idea of an intelligence-enhancing machine into reality.

Engelbart has gone down in Silicon Valley folklore as the man responsible for what became known as 'The Mother of All Demos'. In the winter of 1968, at a computing conference in San Francisco, Engelbart's team rigged up a giant screen to demonstrate their new system. That system represented the first real-life model of what personal computing could look like. It included a workstation with the first ever mouse, a graphic display like a modern computer screen, a keyset, and a working form of hypertext. Whenever you see a tech CEO standing on a stage in front of a giant screen demonstrating their new wonder technology, they're trying to recapture the magic of Engelbart's first demo over 50 years ago.

Engelbart's inventions were the product of a project to augment human intellect. Like Bush and Ashby, Engelbart began with the problem of complexity, and the fear that human problem-solving was not cut out to address the increasingly urgent challenges of the modern world. The way to improve that problem-solving ability, he believed, was to find new ways for man and machine to work together, to give people access to the best technological aids, and to develop new ways to use these technologies to tackle problems.

This solution to the problem of modern complexity had, according to Engelbart, come to him as a vision one day in December 1950, when, at the age of twenty-five, he was casting around for what to do with his life. An image, he recounted later, flashed into his mind of a person sitting in front of a big screen filled with symbols, pushing buttons, knobs and switches, not unlike Bush's vision of the Memex. The person was surrounded by colleagues sitting at similar workstations, working collaboratively through the same computer system. In working with these machines and with each other, they were evolving new skills and ways of thinking.[24] The vision might sound mundane today, but in 1950 there were only about a dozen computers in the whole world, and they were almost all being used to do mathematical calculations for the military. Engelbart's ideas were genuinely revolutionary.

By the end of the 1950s he had developed this vision into a project at the prestigious Stanford Research Institute in Palo Alto. With funding from the US Air Force, NASA and the Department of Defense research arm – the Advanced Research Projects Agency (ARPA) – Engelbart and his team would spend over a decade trying to bring his vision of the modern computing workstation to life.

Some of their ideas were a little on the eccentric side. An early NASA-funded programme, for example, tried to develop a 'nose pointer' to control a cursor on a monitor. This effectively involved strapping a laser pointer joystick to a hard hat and aiming it at a screen. Perhaps unsurprisingly, users found the device difficult to control and complained that it gave them a sore neck.[25] More promising was a knee-controller, where the user could manipulate a cursor on the screen by moving their knee from side to side or up and down, their movements registered by a device strapped to the bottom of a desk. New computer users actually found this very intuitive, but more experienced ones complained that it was too slow. Both of these ideas were motivated by the desire to leave users' hands free to manipulate other controls. Eventually they were abandoned in favour of the mouse, the product of another NASA-funded project which

Engelbart designs. English, Engelbart and Berman, 'Display-Selection Techniques for Text Manipulation', IEEE Transactions on Human Factors in Electronics, 8:1 (March 1967), 5–14. Top: The online system work station showing the CRT display, keyboard, push buttons and mouse. Bottom: Knee control bug-positioning device

turned the light guns used in radar systems and the light pens developed by other computer researchers into a more user-friendly form.

Alongside the mouse, Engelbart's team developed new ways to display and modify text on a screen. They explored tools that would allow users to retrieve deleted material. They produced systems to integrate offline and online information. They found a way to display video camera footage on a computer monitor, and developed a form of windowed user interface. They thought deeply about the ergonomics of workstations, including a 'yoga' workstation where the user would sit on cushions on the floor. And they were one of the first two nodes connected to the ARPA Network (ARPANET), the forerunner of the internet, developing their own electronic mail system.

Underpinning these practical technologies was a much broader vision of what they were trying to achieve. The ultimate goal of the project, Engelbart stated modestly, was to 'increase significantly the intellectual effectiveness of human beings'.[26] This was a prerequisite for finding solutions to the urgent, complex problems facing mankind. If it could be achieved, he argued, its impact would be comparable to splitting the atom, conquering space or curing cancer. 'The survival of man,' he wrote to NASA in 1967, 'may even depend on it.'[27] (Not a bad line to use when you're asking for more research funding.)

This idea – that machines could amplify human intelligence – built on the arguments of Vannevar Bush and Ross Ashby. Like Ashby, Engelbart was less concerned with defining intelligence than with finding ways to help people solve problems. He was talking not about enhancing native human intelligence but about amplifying that intelligence by finding ways to help humans make effective use of new computing tools. He viewed the computer as a kind of prosthesis, a device that could extend human intelligence beyond the body. Although he collaborated with colleagues working on the early development of AI, his vision was crucially different. His goal was not to create a new kind of machine intelligence independent of the human brain. It was to use machines to *augment* human intelligence.

Despite some of the incredible developments that emerged from Engelbart's team in Stanford, the project eventually collapsed in the early 1970s. In many ways it was a product of the California counter-culture of the era; Engelbart's vision of unleashing human abilities tapped into the language about human potential and self-actualising that were swirling around places like Esalen. But the strength of his personal vision resulted in a top-down management style which ran up against the anti-hierarchical ethos of the era. Rifts in the team emerged after many of its members began to attend a controversial, Zen-infused personal development programme which boomed in popularity around San Francisco in the early 1970s.[28] Eventually most of the key figures left, many moving on to the nearby Xerox Palo Alto Research Center (PARC) and later to organisations like Apple. It was in these places that modern personal computing was born, often drawing directly from the work of Engelbart's original project.

It's worth pausing to reflect on some of the biases, limitations and prejudices hidden within all of these visions of computer-aided intelligence augmentation. All were to some degree elitist, built around the idea of what Engelbart called the 'intelligence worker' – what would later come to be more commonly referred to as the 'knowledge worker' – whose jobs revolved around decision-making, problem-solving and mastery of written information. Engelbart's project focused even more narrowly on the computer programmer as representative of the knowledge worker in general, but argued that the same technologies could be applied to the work of scientists, diplomats, executives and lawyers. His vision of the co-evolution of human and machine was one in which the human computer-user was a highly trained expert. But it had little to say about the role of non-experts, or the impact on all the people who weren't 'intelligence workers'. Even some of his colleagues criticised what they saw as an anti-democratic vision of the future of computing.[29]

Indeed, computing visionaries like Engelbart and Ashby recognised that these technologies might end up harming ordinary people. They often referred to the machines of the Industrial Revolution, which had automated away a lot of manual work. Now they argued that computers would be able to automate much of the routine brain work of a more educated segment of the labour market. The only meaningful work that would remain for these people would be to support and enable the small proportion of elite knowledge workers whose intelligence was being augmented by machines. Like many computing pioneers, Vannevar Bush's description of this process had been overtly gendered, describing computers being fed information and instructions by 'a whole roomful of girls armed with simple keyboard punches'.[30] Whether explicitly or implicitly, the humans whose intelligence was going to be amplified were imagined as men working in elite jobs.

Cognitive enhancement

Now, you may be reading all this thinking how quaint it seems – all these funny stories about people trying to boost intelligence with their computer consoles and their naked yoga retreats and their baby-rearing classes. What a strange bunch our 20th-century forebears were!

If that is indeed what you've been thinking, then I'm afraid I've got some bad news. Because these ideas never went away. They have been dragged, alive and kicking, into the 21st century. Intelligence augmentation is very much still – as the kids say – a thing.

The term most commonly used to refer to these ideas today is 'cognitive enhancement'. It's a term that can refer to lots of different things: improving nutrition for children and pregnant women to boost cognitive function; digital teaching and 'brain training' apps; new techniques for communication and knowledge sharing; and even direct electrical stimulation of the brain. But the thing it's most commonly associated with today is the use of so-called smart drugs.

Smart drugs are not a new phenomenon. Substances like coffee have been recognised for centuries as mental stimulants. And when Aldous Huxley spoke about unlocking human potential in the 1950s and 1960s, one of the ways he thought that could happen was through psychedelic drugs like mescaline and LSD.

The modern history of smart drugs really began in the 1880s, when scientists managed to identify and extract the alkaloid cocaine, the active ingredient of the coca plant, which had been used in Peru for centuries to enhance stamina. In the wild west that was the pharmaceutical industry of the late 19th and early 20th centuries, cocaine was aggressively marketed as a catch-all cure for everything from toothache to depression. It quickly became valued as a way to improve mental performance, and powers of concentration and attention. Sigmund Freud was a particular fan, using it liberally as a way of raising his 'intellectual and bodily vigor' as he juggled the challenges of raising a family and building a career in fin-de-siècle Vienna.[31]

Cocaine's popularity as a smart drug declined sharply in the 1920s. This was partly because the problem of addiction became increasingly obvious. But it was also linked to the increasing popularity of the drug among the poor and socially marginalised. Luckily, this coincided with a new discovery by a young Los Angeles-based chemist called Gordon Alles. Alles was trying to develop medication for allergies and asthma. In 1929 he managed to synthesise a new chemical which came to be known as amphetamine. Amphetamines didn't have much effect on the allergies of Alles's patients, but they did seem to perk them up a bit. For good measure, the following year Alles also managed to synthesise the amphetamine derivatives which came to be known as ecstasy and MDMA.

The patents for these new products were swiftly snapped up by drug companies. Just as with cocaine, new amphetamine brands like Benzedrine were sold as a wonder treatment for a wide range of problems – from mood disorder and nasal congestion to narcolepsy and obesity. But researchers quickly began to notice that they also seemed to boost mental performance. Scientists in London reported that amphetamine use boosted scores on intelligence tests. Others working in a Pennsylvania hospital dosed up some of their nursing colleagues and found they showed improved task performance, consistency and alertness. By the late 1930s the *New York Times* was warning that students had started using them as 'brain fuel' to get them through exam season.[32]

Amphetamines really hit the big time in the Second World War. The Nazis found that they boosted the stamina and aggressiveness of troops tasked with implementing Hitler's *blitzkrieg*, handing out 35 million tablets to troops between April and June of 1940 alone. Not to be outdone, the British and Americans also began peddling them as all-purpose boosters of morale, cures for 'war neurosis' and 'combat fatigue', and pick-me-ups for pilots and bomber crews. Churchill recommended them to British commandos sent into Greece, and Montgomery gave out 100,000 tablets to improve the 'fighting spirit' of his troops before the Battle of El Alamein. The British military used 72 million amphetamine tablets over the course of the war. The US military gave them to around 12 million servicemen, a quarter of the entire adult male population.[33]

Not surprisingly, lots of people emerged from the war hooked. Amphetamine prescriptions continued to grow, popular among everyone from suburban housewives to beatnik writers like Jack Kerouac, who took them to boost his creativity. As with cocaine, a crackdown followed in the 1970s, when addiction became an obvious public health problem and amphetamines became increasingly associated with the 'wrong' sort of user. But that doesn't mean they went away. Alongside the continued illegal use of drugs like ecstasy and crystal meth, amphetamines continued to be used, perfectly legally, as treatment for sleeping and hyperactivity disorders. Amphetamines like Adderall, and amphetamine-related stimulants like Ritalin, were used particularly widely to treat ADHD, diagnoses of which have skyrocketed since the 1970s.

And this brings us back full circle to the modern practice of 'cognitive enhancement'. Because the way these drugs help to boost memory, concentration, focus and attention for the 'unwell' also appeals to the 'well' who are looking to boost their mental performance. A 2006 study found that 16 per cent of college students in the US had used drugs for cognitive enhancement.[34] Drugs like Ritalin, they reported, improved their attention, boosted their grades and allowed them to combine study with socialising. Other surveys suggested they had been used by one in five academics.*

There's been a lot of debate about the rights and wrongs of these drugs. Critics have warned about the obvious risk of unpleasant side effects. They've warned the drugs may encourage risky behaviour by people like long-distance drivers or doctors using them to cope with tiredness. And they've highlighted the unfair advantage these drugs potentially give those taking competitive academic exams, such as for admission to elite universities.

But there are a lot of people, and indeed a lot of very respectable organisations, who've argued that cognitive enhancement could be a good thing. The British Medical Association argued in 2007 that we all benefit from the intelligence of skilled scientists, doctors and (apparently) politicians, so it would surely be better for society if there were more of these people.[35] Assuming universal access to brain-boosting technologies, they could be used to raise everyone's cognitive level, giving Britain a competitive advantage in international trade. A 2012 report by the British Academy, the Royal Society and other professional bodies argued that cognitive enhancement would 'enable more people to work at their full biological capacity', raising standards and increasing workplace diversity.[36]

---

* I have to say this isn't something I've personally come across in university life, but it may be that my colleagues are just good at hiding their rampant brain-boosting drug habits.

These kinds of arguments have a lot in common with the ideas about intelligence and national development that underpinned the Ministry for the Development of Intelligence in Venezuela. But they also tap into the notion of 'transhumanism', a philosophy which has risen to prominence in recent years. Transhumanism – alongside the different but related idea of 'posthumanism' – first emerged in the 1980s. Broadly speaking, it revolves around a set of beliefs about the future of humanity. Transhumanists tend to believe that technologies will emerge in the future that will allow us to hugely enhance human abilities and potential, including our intelligence, our longevity and our happiness. They believe that people should be free to explore the extension of these limits now, whether through the development of AI and projects to merge biological and digital intelligence or efforts to cheat the ageing process through ideas like cryogenics or 'uploading' minds to a computer. They think about the long-term human future, worrying about existential risks to our species and exploring ways for us to escape the confines of earth. Because of this long-term focus they believe that we have a greater responsibility towards the well-being of future generations than we normally take into account.

These ideas first began to spread through magazines, early internet message boards and email lists from the late 1980s. International transhumanist associations and conferences emerged from the mid-1990s, and the first transhumanist declaration was published in 1998. It became particularly associated with the computing, tech and AI communities. Marvin Minsky, for example, who had been pioneering the development of AI since the 1950s, began writing in the 1990s about extending human life and replacing brains with computers. By the 2000s it became institutionally rooted in places like the University of Oxford's Future of Humanity Institute, putting down deep roots in Silicon Valley and expanding its reach through the vocal support of celebrity tech entrepreneurs.[37]

Transhumanism looks towards the future but is rooted in the past – specifically the history of ideas about the improvement of the human species and the augmentation of intelligence. The person who first coined the term 'transhumanism' in the 1950s was the British biologist Julian Huxley, the younger brother of Aldous Huxley who we came across earlier in this chapter.[38] Huxley, perhaps best known as the first director-general of UNESCO, was one of the leading liberal eugenicists of the mid-20th century. An anti-racist who rejected the coercive eugenics that came to be associated with Nazi Germany, he nonetheless believed that the new science of genetics needed to be actively used to improve the genetic quality of the human race. His idea of transhumanism stemmed from his belief that mankind was entering a new evolutionary era in which it would be able to shape its own genetic destiny for the first time.

Huxley's eugenics was rooted in ideas about the importance of intelligence and the need to augment it. Although he rejected the idea of fundamental racial differences in intelligence, he was clear that individuals were born with very different intellectual gifts, and that progress was driven by the most intelligent.[39] The new science of genetics, he hoped, would open up new ways to improve the intelligence of everyone, transforming intelligence, even genius, into a birthright.

This vision, of humanity transformed by new technologies and techniques to magnify intelligence, is a thread that has run throughout modern history. It's something that links together early 20th-century eugenics, the post-war movements for human potential, the attempts to create techniques and technologies to help us think more effectively, and the transhumanism of the early 21st century.

And it all rests on the broader history of the way we came to think about intelligence. To believe that intelligence *could* and *should* be augmented, people first had to come to believe that intelligence existed as a single, coherent *thing*, and that that thing was the key to human progress. As we've seen, these were fundamentally modern ideas, ideas that emerged in the late 19th century and became embedded in modern society, particularly Western society in the early 20th century. From the 1960s, these ideas had become largely unchallenged. When combined with the resurgent belief that intelligence was shaped more by the environment than by heredity – nurture rather than nature – the next logical step was to ask how this key commodity could be multiplied.

This was an idea that was sometimes framed in an egalitarian way – the intelligence of *everyone* could and should be augmented, through better parenting, different kinds of education, access to new drugs and so on. But it was impossible to separate ideas about intelligence from ideas about inequality. Often it was the brainpower of a narrow elite that was assumed to matter most. If it was this cognitive elite which drove progress, so the thinking went, the way to truly make an impact would be to further augment their intelligence. The normal intelligence levels of everyone else could be disregarded, and the gap between the talented tenth and the rest could be allowed to grow.

It was the persistence of these ideas about intelligence and inequality – even in an era that emphasised the environmental factors shaping intelligence levels – which helped to drive a backlash against intelligence towards the end of the 20th century. This backlash partially reversed the triumphant march of intelligence that had begun at the end of the 19th century, seeking to overturn some of the ideas about intelligence that had become embedded in modern society, and to develop new, less restrictive ways to think about human mental ability.

9

Resisting and Rethinking Intelligence

'The closed world of intelligence is being opened up.'
Howard Gardner

IN APRIL 1969 HUNDREDS of people gathered outside Haringey civic centre in Wood Green, North London. Most were Black parents and young people from groups representing the local West Indian community, alongside Greek Cypriot parents and other supporters. Ranged against them was a counter-demonstration from the fascist National Front, carrying banners reading 'Start Repatriation' and 'Send them Back'. Inside the centre over 100 people had crammed into the public gallery to hear a council debate. The disruption they caused led to 30 people being expelled from the chamber, and forced the mayor to walk out twice.

The protests were the culmination of months of campaigning begun in response to a report published by a man named Alfred Doulton. Doulton was the headmaster

Haringey anti-banding protest (1969). Source: George Padmore archive

of an elite public school in North London and the vice-chairman of Haringey's education committee. His report aimed to address the 'problem' of immigrants in Haringey's schools. The borough had some of the highest immigration levels in the country, with around 30 per cent of children from immigrant backgrounds, reaching 70 per cent in some primary schools. Doulton felt that the council had to take this into account in its education system.

But his report included one particularly notorious phrase that provoked the ire of the local Black community, and which sparked the protests that came a few months later. 'On a rough calculation,' he wrote, 'about half the immigrants will be West Indians at 7 of the 11 schools [in the borough], the significance of this being the general recognition that their I.Q.'s work out below their English contemporaries.'[1]

In short, Doulton's argument was this: Haringey was going through a process of comprehensivisation – replacing the previous system of selective grammar schools and non-selective secondary moderns with new comprehensive schools for children of all academic abilities. The children of immigrants, particularly Black immigrants from the Caribbean, were going to form the majority of pupils at schools in the neighbourhoods where they were most concentrated. Given that, as Doulton believed, these children were inherently less intelligent than their White English counterparts, these schools would inevitably have lower academic standards and would be unable to support the Sixth Forms which would enable their pupils to study for A-levels. They would, in the words of another council report, be 'ghetto' schools, where low standards and social divisions would become baked in.[2]

The solution Doulton proposed to this supposed problem was 'banding' – testing the intelligence of all primary school children so that they could be arranged into bands of academic ability, then ensuring that an equal number of children in each band were distributed to each school across the borough. Once in those schools, children would then be divided into 'streams' according to their academic ability, and the different streams would be educated separately. The assumption was that Black children would cluster in the lower bands, would therefore be directed away from the schools in immigrant-majority neighbourhoods, and would then largely be educated in the lower streams of White-majority schools.

Unsurprisingly, these proposals weren't well received by Haringey's Black community. As soon as the report became public in March 1969, mobilisation against the plan began. The North London West Indian Association (NLWIA) joined the Greek Parents' Association in petitioning the council, denouncing the 'offensive and dangerous remarks' made about their children's intellectual capacities. They argued that the panic over 'the education of immigrants' was merely glossing over the real problems of an inadequate and underfunded education system. These problems could not be solved, they argued, without fundamentally changing England's class-stratified educational model.[3] In the meantime, they called for the scrapping of the banding policies, more nursery and preschool provision, the hiring of more Black teachers, and the teaching of Caribbean history and literature as part of the curriculum.

Despite these concerns, and in the face of the mass protests gathered outside, the council meeting in April approved the banding proposals. But mobilisation against them continued. The High Commissioners of Jamaica, Barbados and other Caribbean states in London tried to persuade Doulton about the unreliability of IQ tests, particularly tests designed for British-born children being given to children who might only recently have arrived in the country.⁴ An article in Darcus Howe's *Black Dimension* newspaper warned that if the policy was accepted in Haringey then measures to disperse Black children would be implemented across the country.⁵ 'They claim we are inferior in intelligence and are lowering education standards,' declared one campaign poster, 'What rubbish!'⁶

Ultimately these protests succeeded in delaying the implementation of the banding policy. When the next local elections were held, in May 1970, the incumbent Conservatives were ousted and the new Labour administration dropped the proposals. But the legacy of the protests lived on, marking a crucial step in the development of the Black Education Movement in Britain. Many of those who fought the Haringey banding proposals would go on to lead the fight against the high proportion of Black children being sent to schools for the 'educationally sub-normal', and for the creation of Black supplementary schools.⁷

The Haringey protests were a symptom of a broader change. After the Second World War, IQ testing and popular faith in intelligence had reached their zenith. The rigid vision of IQ developed by intelligence scientists had spread around the world and had embedded itself into popular culture. Intelligence testing had consolidated its hold over schools and workplaces. This was the era that witnessed the birth of Mensa and the idea of AI.

But discontent was brewing. We know there were people earlier in the century who had criticised modern ideas about intelligence and intelligence testing, people such as the American journalist Walter Lippmann, who had publicly challenged Lewis Terman in the 1920s. Plenty of other groups had criticised the way ideas about intelligence had been used to reinforce inequality. Feminists across the 20th century, for example, had challenged assumptions about women's inferior intelligence. Aside from some of the earliest researchers, most intelligence scientists had quickly abandoned the idea that there were major intelligence differences between men and women. But the idea of women's intellectual inferiority has deep historical roots which feminist thinkers and activists tried to tackle. In the late 1940s, for example, Simone de Beauvoir challenged the tradition of scientific research into 'psychophysiological' gender differences, such as the attempt to demonstrate different levels of intelligence by measuring brain size. She also denounced the effect of the inferiority complex bred by a culture which defined women by what they lacked, by the ways they fell short of the male ideal. This complex, de Beauvoir argued, weighed heavily on the modern woman's intellectual accomplishment. It encouraged her to hide her abilities, culture and intellectual achievement from men. It convinced her that her capacities genuinely were limited. And this in turn caused her to repress her intellectual, educational and professional ambitions. '[B]y resigning herself to this inequality,' de Beauvoir lamented, 'she enhances it.'[8]

By the 1960s, though, these kinds of criticisms were becoming mainstream. In an egalitarian decade impatient with social hierarchies, the idea that a rigid notion of intelligence should determine an individual's life chances was coming under increasing attack. Anti-racism campaigners like those in Haringey were no longer prepared to let pseudo-scientific arguments about intelligence be used to legitimise discrimination. Parents of all races and classes were increasingly unwilling to accept hierarchical education systems that condemned the majority of children to substandard schooling. These egalitarian trends were met with a backlash, including attempts to revive theories about racial differences in intelligence. But in response, a new generation of psychologists and activists began to develop more inclusive theories about the nature of intelligence and the human mind.

The late 20th century, then, was an era in which ideas about intelligence were being *resisted* and *rethought*.

Equal education

The area where ideas about intelligence were most obviously challenged during this period was education.

This was very different from the situation in the early 20th century. As we saw in Chapters 3 and 4, intelligence and intelligence testing had become embedded in education systems after the First World War. Education had expanded rapidly in most countries during the first half of the 20th century, and this 'mass education', as it was commonly called, posed new logistical challenges for governments. Many feared that it would threaten the existing social order.

In response, almost all countries developed educational systems that were divided and hierarchical. Children were separated into different educational streams or tracks. A minority were given access to an academic education through which they could gain entry to universities, which were still confined to a narrow elite. These children would go on to form part of the managerial class, or would provide the next generation of teachers to keep the education system going. The rest were channelled into various forms of practical, technical or vocational education, designed to produce the workers of the future. The decision to allocate children to these different streams rested on some kind of judgement about their intelligence, and was often based explicitly on an intelligence test.

This was more or less what happened in England and Wales (the Scottish education system was, and remains, different). Intelligence tests were first used in British schools after the First World War, mainly to make decisions about which children would be selected for the more academic grammar schools. Supporters of intelligence tests saw them as a tool of equality and progress. Instead of access to these elite schools depending on the wealth and class background of a child's family, it could now be based on an objective measure of their minds. The tests were meritocratic tools, offering equality of opportunity and the promise of social mobility where neither of these things had existed in the past.

This educational model became further embedded after the Second World War. Under the so-called 'tripartite' system, all children were educated together in elementary schools up until the age of 11 (although many of these schools separated their pupils into different streams). They then sat the 11+ exam, which was modelled on an intelligence test. On the basis of this exam, children would be assigned to either grammar schools, technical schools or secondary modern schools.

We tend to think of the post-war years in Britain, when the Labour government built the welfare state and created the NHS, as a period of egalitarianism. But, as these education reforms suggest, the desire for genuine equality was much less powerful than ideas about meritocracy and equal opportunities. 'British people want to make excursions up the side of the social pyramid,' as one contemporary observer put it. 'They do not so far consider that there is any law of nature or justice which should make it certain that they can do so, though there is a growing feeling that their children should be able to do so, if they have the ability and wish.'[9]

But the tripartite educational model, and the ideas that underpinned it, came under increasing attack from the late 1950s. Officials had claimed that there was parity of esteem between each of the three types of English schools – that although children might be separated into different educational tracks, it was wrong to think about any one of these tracks as being better or worse than the others. There were a couple of problems with this claim. First, just because someone in Whitehall stated that all types of school were equally esteemed, it didn't automatically mean that everyone in the country would agree to think the same way. Parents were well aware that employers, universities and society at large continued to see grammar schools as inherently superior to technical or secondary modern schools. The second problem was that this hierarchy of esteem was made patently obvious through government policy. Grammar schools tended to be well-funded and resourced. Technical and secondary modern schools, particularly those in poorer areas, were chronically underfunded, their pupils consigned to leaking buildings and condemned to bulging class sizes.

All of this prompted a groundswell of popular opposition to division, hierarchy and selection in the education system. Parents increasingly demanded 'grammar schools for all', reflecting a growing popular faith in the power of education and an unwillingness to see most children condemned to substandard schools.[10] This popular discontent boosted support for a radically new approach to education. As the 1960s approached, more and more people began to argue that the tripartite system should be replaced by comprehensive schools in which all children would be educated under the same roof. Comprehensives, they argued, would bring true equality to the education system for the first time – equal resources, equal opportunities, equal treatment.

Debates over intelligence and intelligence testing were key to the rise of comprehensive schools. In the immediate aftermath of the Second World War there had been little challenge to the use of intelligence testing in the school equality of outcome. The commitment to education and social mobility had deep roots in the British left, particularly within the elitist Fabian movement. This faith in meritocracy was reflected in the fact that many of the new comprehensive schools continued to stream their pupils, often relying on the same kind of tests that had been used for the 11+. Nevertheless, the shift from the tripartite to the comprehensive school system represented a genuine attempt to break down educational hierarchies, and a rejection of intelligence as a basis for educational inequality.

Similar ideas were reshaping higher education at the same time. Some of the same experts who had been promoting comprehensive schools in Britain were involved in international debates about universities and the so-called 'pool of ability'. These debates raised the question of how many young people had the intellectual abilities needed to genuinely benefit from higher education.

Universities had traditionally been restricted to very narrow elites. There had been some expansion of higher education in the first half of the 20th century, but politicians and professors had traditionally assumed that only a very small proportion of the population were suited to higher education. This began to change after the Second World War. The United States, which had already led the way in the expansion of higher education before the war, massively increased access to college programmes. This was driven in part by the G.I. Bill, which funded college education for military veterans. But it also reflected increased demand for skilled workers in the booming post-war economy.

This sudden expansion of the university population prompted education experts to question whether their previous assumptions about the pool of ability had been too low. Perhaps a much greater proportion of the public could benefit from higher education than had previously been thought? At this stage, though, the pace of change was still quite limited. Up until the late 1950s, experts were estimating that only around 15 per cent of young people could usefully be sent to university.[13]

Much more radical change came with the massive expansion of university education in the 1960s. This was a global phenomenon. More than 200 new universities were built around the world during the decade.[14] Western, communist and post-colonial states all wanted to ensure they were producing enough scientific 'manpower' (in the terminology of the time) to drive economic growth. And educational experts were going much further in traditional assumptions about access to higher education. Rather than just debating the size of the pool of ability, these experts began to challenge the concept altogether. Ability, they now argued, was not a singular aptitude, and there was no good scientific evidence that the amount of human intelligence in a population was genetically fixed. Given the right social and cultural conditions, there was no reason why everyone shouldn't be able to benefit from advanced education.[15]

The proportion of young people attending university and the precise way they were selected varied widely between countries. But in almost all parts of the world, access was being opened up and previous barriers broken down. Many of the new constitutions of post-war countries guaranteed university access to all who met the entrance criteria. In socialist Yugoslavia, higher education was made available to people without school leaving certificates as long as they could demonstrate relevant practical experience. In Zambia, the entry requirements for university were lowered from A-level to O-level. In Norway, universities began selecting students by drawing lots.[16] In Britain, the Robbins Report of 1963 recommended the expansion of student numbers by half a million. The Open University was founded a few years later to radically widen access to higher education.

But it was once again the United States that led the way. In certain parts of the country there was a move towards nearly universal open admission to college – effectively a promise that any young person could access higher education as long as they met very limited entry requirements. This was still being driven in part by the demands of a booming post-war economy. But it was also linked to Lyndon Johnson's Great Society reforms, and to concerns about racial inequalities in higher education which were receiving more attention with the rise of the Civil Rights movement.

By the late 1960s, 40 per cent of college-age Americans were enrolled in full-time degree programmes, far ahead of comparable countries in western Europe. In places like California around 80 per cent of high school graduates went to college. The new California 'Master Plan' for education had decreed that any child in the state could attend a two-year community college programme. Above that level there was some stratification – certain colleges were open to students in the top third of their high school class, others to those in the top eighth – but tuition remained free and there was no selection process beyond the basic grade and subject requirements.[17] Even elite colleges like Berkeley only required students to meet the basic entry criteria. Many colleges elsewhere in the country were effectively moving towards open admissions. Institutions like Kansas State University, for example, admitted any state residents with a high school diploma. Similar policies were introduced by CUNY in New York, where open admissions dramatically increased the proportion of Black and Puerto Rican students entering college.[18]

These radical changes provoked some backlash. Critics like the English novelist Kingsley Amis argued that, when it came to universities, 'more will mean worse'.[19] Others pointed to the high drop-out rates in the expanded US colleges, and the lack of motivation among students who merely drifted into college on the back of remedial programmes and generous grants – a 'semidrafted army of students', as one British vice-chancellor put it.[20] The big expansion of student numbers wasn't always met by increased investment, meaning overcrowded lecture halls, poor teaching and crumbling accommodation – some of the factors that fed into the global wave of student revolts that broke out in 1968. Indeed, many of these criticisms came from students themselves. While some on the '68 barricades had called for an end to educational elitism, over 80 per cent of students at the prestigious French Institutes of Technology said they supported strict academic selection.[21]

These criticisms fed a backlash against the egalitarian trend in education during the 1970s and 1980s. Many universities and colleges were forced to roll back their open door admissions policies. The rapid expansion of student numbers began to slow in some countries. In Britain, there was louder criticism of the shift to comprehensive schools. A series of high-profile *Black Papers on Education* were published from the late 1960s – articles which criticised many of the progressive educational reforms of recent decades. They included work from intelligence scientists like Cyril Burt and Hans Eysenck (more on him later), complaining that supporters of comprehensive schools were refusing to confront the scientific reality of biological differences between children's intelligence levels. One psychologist, Richard Lynn – who would go on to become one of the most notorious defenders of race science and IQ – argued that 'slum dwellers' had low innate intelligence, and that comprehensive schools would lead to 'a new dark age in which the envy, malice and philistinism of the masses' would destroy Western culture.[22]

There were limits to this backlash. In Britain, the comprehensive system remained in place in most parts of the country. Across the rest of the world, increasing numbers of children continued to complete secondary schooling and go to university or college. But the optimism of the great era of egalitarian education reform had come to an end. Despite the best efforts of reformers like Brian Simon, the idea of intelligence would remain central to global debates about education in the late 20th century.

Race, intelligence and resistance

Let's return to the 1969 Haringey protests which began this chapter, and which took place in the context of the shift to comprehensive schools just discussed. These protests about IQ and Black education didn't happen in a vacuum. The British government, previously keen on Commonwealth migration to fill labour shortages, placed new restrictions on non-White immigrants from the 1960s. This coincided with a moral panic over the supposed failure of migrant groups to assimilate into British society.

The education system was seen as both a symbol of and a possible solution for these problems. Immigrant-majority schools were regarded as a problem in their own right, epitomising the apparent tendency of Black and Asian communities to separate themselves from White British society and culture. If these children could be dispersed across White-majority schools, policy-makers came to believe, they would assimilate in ways their parents had failed to do. From the mid-1960s, the British government began recommending quotas for immigrant children in individual schools, and introduced bussing policies in areas like Southall in West London. Bussing, as with Haringey's banding policy, was intended to disperse Asian children to schools away from their local community to ensure that the proportion of immigrant children in any one school wasn't 'unduly high'.[23]

At the heart of the supposed problem of immigrant education was the idea that Black children were educationally and intellectually inferior. The 1950s and 1960s had seen a scientific turn away from the argument that intelligence was hereditary and that some races were biologically superior to others. Official reports on Black educational underachievement tended to at least gesture towards the damage caused by racial prejudice and discrimination. But psychological and educational research at the time still reinforced ideas of Black inferiority.

It did so by shifting the focus from race to culture. Those researchers who believed that intelligence was shaped more by a person's environment than their genetics could still argue that poor educational performance was a product of a primitive culture or deficient home environments. British researchers, even those who accepted the problem of cultural bias in intelligence tests, argued that intellectual development of Black children was retarded by their knowledge of 'immature languages' like West Indian Creole; by bilingualism; by family instability; by poor health and nutrition; by low cultural levels at home; by lack of parental interest or encouragement; or even by the low level of economic and technical development in the Caribbean.[24] These researchers might not have believed that Black children were born less intelligent than White children. But they could still argue that Black culture had *made* those children less intelligent.

These were the kinds of ideas that were being channelled in Alfred Doulton's notorious Haringey report of 1969. The resistance to Haringey's banding proposals demonstrated the anger that such ideas provoked, and the power of local communities to challenge them. Just two years after those protests, another powerful challenge to these ideas was published by an activist and educator named Bernard Coard.

Coard had been born in Grenada in the 1940s, and had studied in the US before moving to Britain. He had spent a number of years running youth clubs in South London and teaching in 'educationally subnormal', or ESN, schools. These schools were designed to provide support for children with serious learning disabilities. At the youth clubs where he worked, Coard was shocked to discover that a large proportion of the Black children he met had been sent to such schools. Whereas immigrant children made up 17 per cent of pupils in inner London schools, he discovered, they represented 34 per cent of children in ESN schools, the vast majority of Caribbean origin. He set out to discover why this had happened and what impact it had had on the children involved. In 1971, he published his findings in a pamphlet, *How the West Indian Child is Made Educationally Subnormal in the British School System*.[25]

His conclusion was that this was partly a deliberate policy – the government was using ESN schools as a dumping ground for Black students, even those with average or above-average intelligence levels. The heads of these schools themselves reported that a large proportion of their Black pupils had been wrongly placed. But once placed there, it was almost impossible for these children to escape back into mainstream schools. They were condemned to an education designed merely to teach basic social and life skills, rather than to help them reach their full intellectual potential.

But it was also, Coard argued, the result of the prejudices and biases of those working in the education system. Referrals for ESN assessments initially came from teachers. Educational psychologists were then tasked with administering intelligence tests and recommending whether a child should be sent to an ESN school. Many of these teachers and psychologists were openly hostile to Black children. Others were patronising. Almost all had low expectations of what their Black pupils could achieve, rooted in their race and class biases. As Coard described the process:

> Most teachers absorb the brainwashing that everybody else in the society has absorbed – that Black people are inferior, are less intelligent, etc, than White people. Therefore the Black child is expected to do less well in school. The IQ tests which are given to the Black child, with all their cultural bias, give him a low score only too often. The teachers judge the likely ability of the child on the basis of this IQ test. The teacher has, in the form of the IQ test results, what she considers to be 'objective'

confirmation of what everybody in the society is thinking and sometimes saying: that the Black children on average have lower IQs than the White children, and must consequently be expected to do less well in class.[26]

And the worst thing about all this, in Coard's view, was that these attitudes were internalised by Black children themselves. Feeding these children through a system which signalled in so many ways that they were intellectually and culturally inferior ended up irrevocably harming their sense of self-worth, their expectations of what they could achieve, and their intellectual ambition.

Coard wasn't necessarily seeking to challenge the concept of intelligence or the use of IQ tests to identify those with special educational needs. But he was clear about the problematic way these tests were designed and used. Black, predominantly working-class children, he argued, were being given tests built around White middle-class vocabulary and ideas. These tests were almost always administered by White psychologists, something researchers had shown tended to lower the scores of Black test-takers. Irreversible decisions about a child's education were then being made based on their performance on a single test on a single day. No allowance was made for their personal experiences of formal tests, their recent experiences with the British school system or their emotional state.

Coard's work helped to further galvanise the early Black education movement in the UK which had already begun to coalesce around the 1969 Haringey campaigns.[27] When he began to present his research, many Black parents whose children had been sent to ESN schools were shocked. They had believed what they had been told by schools and teachers – that these were special schools that would provide their children with the extra resources they needed to eventually return to and thrive in mainstream education. A coalition of Black community groups supported and funded the publication of Coard's pamphlet. It was the first work produced by the Caribbean Education and Community Workers Association (CECWA), a new campaign group that included many of those who had been involved in the Haringey campaign. Over the following years, CECWA would be at the forefront of the fight to identify and free children who had been wrongly placed in ESN schools. They also led the creation of Black supplementary schools that were set up around the country to provide the kind of education and support that Black children so often lacked in mainstream schools.

These debates about Black education in Britain were part of a transatlantic controversy over race and intelligence in the late 1960s and 1970s. The fight against Nazism during the Second World War had helped to discredit the kind of open eugenics and racial science that had been embraced by many intelligence researchers in the early 20th century. As the Civil Rights and Black Power movements gained strength in the 1960s, many believed that the science of race and intelligence had been consigned to history. But the truth was that it had never truly disappeared.

One of the events that inadvertently contributed to its revival was the 1954 Brown vs Board of Education ruling in the United States, which declared racial segregation in American education unlawful. In the years that followed, White opponents sought to challenge the ruling by transforming arguments about racial justice into ones about race and intelligence. A school district in Georgia, for example, argued that the racial segregation in its schools was simply the result of intelligence testing which had objectively sorted Black and White children into different groups.[28] Segregation, they claimed, should be allowed if it was grounded on scientific evidence about intelligence differences between races.

Arguments like this received support from a number of White psychologists, who launched attacks on the environmentalist consensus that had emerged in the field of intelligence science. These attacks were then used by conservative critics of US education policies. During the 1960s, the Johnson administration had doubled federal spending on public schools. This was part of its 'War on Poverty', and was influenced by the Civil Rights-era desire to promote equal opportunities. It focused in particular on so-called 'compensatory' education programmes. These were aimed at boosting the achievement of the most disadvantaged children, offering enrichment programmes like summer schools and preschool childcare. Supporters of these programmes sometimes argued that they would directly boost children's intelligence. Even those who didn't make this argument believed that providing a better environment for children would boost their learning. But Conservative critics marshalled the assessments of critical psychologists to claim that such programmes simply wouldn't work. When Richard Nixon took over the presidency in 1969, he threatened to defund the compensatory education programmes.

In the same year, an educational psychologist at Berkeley named Arthur Jensen was asked by the editors of the *Harvard Educational Review* to respond to the question, 'How much can we boost IQ and scholastic achievement?' Jensen's response took the form of a deeply technical article, over 100 pages long, setting out his beliefs about the nature of intelligence, the factors that shaped it, and how they were relevant to the field of education. Jensen, a dry and apparently mild-mannered professor with almost no public profile, was an unlikely controversialist. But his article sparked a huge international scandal, launching a new front in the culture wars and a new era of public debates about race and intelligence.

The article argued that compensatory education – the attempts to raise educational achievement among the most disadvantaged – had failed. It had failed because it was based on two errors: that almost all children possessed roughly the same amount of intelligence; and that educational underperformance was caused by social deprivation. In reality, Jensen argued, academic performance was rooted in IQ, which varied greatly between individuals for largely hereditary reasons. Each child was unique, and that was just a fact of nature. This meant that compensatory education hadn't just failed – it was *destined* to fail. The only way to really improve education was to recognise the vast differences in the natural intelligence of children, to abandon the goal of uniformly raising everyone to the same academic level, and instead to pursue different educational approaches and aims for children of different abilities.

These ideas were controversial enough on their own. But what really launched the debate about Jensen's article into the public sphere were the sections on race and intelligence. The data, Jensen argued, showed that Black Americans scored lower on both academic tests and IQ tests than White Americans. While he believed there wasn't enough evidence to prove why this was the case, he argued that the evidence that did exist pointed more to genetic causes than to environmental ones. No one had yet been able to prove, he claimed, that there was any way that the average intellectual ability of Black American children could be raised to the average level of White American children.[29]

If some of these ideas sound familiar, it may be because they closely echoed those of William Shockley, the inventor of the transistor turned racist agitator we came across in Chapter 7. Jensen had met Shockley in 1968 during a fellowship at Stanford, and remained close to him for the remainder of Shockley's life.[30] Jensen's article touched on many of Shockley's favourite themes, including the need for 'no holds barred' research into racial differences in intelligence, and the danger of declining IQ levels in modern society. Shockley was so enthused by Jensen's article that, at great personal expense, he sent a copy to every single member of the National Academy of Sciences, hoping it would boost his campaign for a new research agenda into the science of race and intelligence.

Jensen had lit the touchpaper. Now it was the turn of one of his mentors across the Atlantic to fan the flames. Hans Eysenck was one of post-war Britain's most respected and well-known psychologists. He was a professor at the prestigious Institute of Psychiatry in South London, a specialist in the study of personality. But he had a particular talent for what today we would call science communication. Since the 1950s he had been writing best-selling books explaining psychology to the general public, appearing regularly in newspapers, and on radio and television. A self-publicist with a big personality, he wasn't afraid to engage with contested topics. His controversial reputation has continued since his death in 1997. In 2019 an enquiry revealed that many of his later papers, which had been part-funded by the tobacco industry and had suggested there was evidence that certain people had 'cancer-prone personalities', were fraudulent.[31]

Jensen and Eysenck had been close since Jensen had come to London to do post-doctoral research in the late 1950s. He had returned regularly to the city ever since, and did so again a few months after his controversial article had been published. Eysenck had followed the debates about Jensen's work with a keen interest, and was eager to do what he could to bolster public support for his friend.

A year after Jensen's visit, in 1971, Eysenck published a book called *Race, Intelligence and Education*, which summarised the recent science of IQ for a general audience and sought to defend Jensen's line of argument.[32] As with Jensen's article, the book discussed racial differences in intelligence and criticised affirmative action policies.

Eysenck had long been interested in intelligence, having published articles on intelligence science in the *Eugenics Review* in the late 1940s. In the years that followed he had also collaborated with the founders of Mensa.[33] Like Jensen, he thought that the psychological community was shying away from the study of race differences for political reasons, something he saw as a betrayal of true scientific principles. And over the preceding years he had contributed to public debates about the Labour government's comprehensive schools policy, motivated, he claimed, by the poor standard of education his own children received in local schools. But unlike Jensen, he had never carried out any of his own research on the science of intelligence. And where Jensen had framed his divisive argument about race and intelligence in relatively moderate and careful language, Eysenck – much more used to writing informally for a wider audience – sprinkled his work with racialised anecdotes and stereotypes. His book was a huge success.

Jensen had launched the controversy about race and intelligence. Eysenck had amplified it across the Atlantic. Back in the United States, one more figure was going to weigh in on the debate. Richard Herrnstein wasn't a relative unknown like Jensen. He wasn't a stranger to intelligence research like Eysenck. He was at the top of his field, the chair of the Harvard Psychology Department. And in 1971 he penned an article in the *Atlantic* entitled simply 'I.Q.'.

The article was in part a defence of IQ and the IQ test, something Herrnstein saw as the greatest accomplishment of the discipline of psychology. But above all it was an argument about intelligence and society. Differences in intelligence, Herrnstein argued, are largely inherited. Intelligence is crucial to professional success in the modern world, and people's income and social standing depend on that professional success. It followed, therefore, that a person's class – their social standing – is dependent at least in part on their genes. People at the top of society, Herrnstein was saying, were where they were because they were better than the rest – more gifted, more intelligent, better fitted to succeed in the modern world.

What's more, Herrnstein believed, intelligence was only going to get more important as time wore on. As society became more prosperous and opportunities more equal, so environmental differences would have less impact on people's life chances. If environmental differences became *less* important, hereditary differences would become *more* important. More equal opportunities wouldn't make society more equal. Instead, the most intelligent would continue to rise to the top, and the least intelligent would sink to the bottom. Modern society, Herrnstein argued, was increasingly becoming a genuine cognitive meritocracy.[34] Michael Young's dystopian vision was coming true.

Together, Jensen, Eysenck and Herrnstein succeeded in catapulting debates about race and intelligence back into the public sphere during the late 1960s and early 1970s. Part of their success lay in the way they tapped into the culture of the time, particularly the conservative backlash against the liberalism of the 1960s. But those same social conditions ensured that their arguments were met with fierce resistance. As the Haringey protests of 1969 had shown, people were no longer willing to allow claims regarding the low intelligence of certain races to pass unchallenged.

Much of the resistance to these arguments came from student groups. Indeed, the controversy about race and intelligence became central to the campus culture wars of the time. In the United States, Students for a Democratic Society organised campaigns against Jensen at Berkeley and Herrnstein at Harvard. In the UK, Eysenck was met with protests at universities where he was invited to speak. When he appeared on the BBC to debate his ideas, student protestors infiltrated the audience and caused so much disruption that producers were forced to invite two of them to join the debate panel. The campaign against him culminated in a protest at the LSE in 1973, organised by the British offshoot of a small Canadian Maoist group, in which Eysenck was assaulted on stage, suffering a small cut on his head and breaking his glasses. Dramatic images of the attack appeared in the national press, bolstering Eysenck's public profile as a champion of race realism in science and a free speech martyr.[35]

Just as with today's campus culture wars, these protests received a huge amount of media and political attention, sparking major debates about free speech and its limits. Student protestors argued that claims about the biological inferiority of certain races amounted to little more than fascism, and that the protection of academic freedom didn't extend to racist pseudo-science. Their campaigns had limited success – neither Jensen, Eysenck nor Herrnstein faced any sanction from their employers. They continued to publish about race and intelligence, and were invited to discuss their ideas at universities around the world. But they, and like-minded colleagues, certainly felt that their freedom was being threatened. A 'Resolution on Scientific Freedom' signed by 50 scientists in 1972 – including Jensen, Eysenck and Herrnstein – modestly drew parallels between their own situation and the Church's persecution of Galileo, or the attacks on Darwin's theory of evolution. 'Today,' they argued, 'a similar suppression, censure, punishment, and defamation are being applied against scientists who emphasize the role of heredity in human behavior.'[36]

Far more damaging to Jensen, Eysenck and Herrnstein than the student protests were the attacks from their scientific peers. Psychologists and educationalists published an avalanche of newspaper articles, scientific papers and books robustly challenging the three men's assumptions, methodology and conclusions. Many of these criticisms situated their work within the longer history of intelligence research in the United States, showing how such research was embedded in the scientific racism of the early 20th century and entangled with the history of forced sterilisation.[37] Bernard Coard, the man who had led the campaign about Black children in British ESN schools, denounced the 'paternalistic stereotyping' of psychologists like Eysenck, showing how they ignored inconvenient research about the non-genetic factors influencing IQ scores, and accusing them of justifying the racist agitation of contemporary politicians like Enoch Powell.[38]

The combined effect of these attacks – from the public protests in places like Haringey to student campaigns and scientific critiques – succeeded in limiting the impact of this race and IQ revival. Despite complaints from people like Jensen and Shockley, research on group differences in intelligence remained largely taboo within mainstream psychology. In the United States, a number of courts and education authorities sought to limit the use of intelligence testing in schools and workplaces, in part because of problems with racial bias. And the wave of publications about race and intelligence seemed to have ended with Herrnstein's *Atlantic* article in 1971.

In fact, it would be over 30 years until the controversy exploded into life again. In September 1994, Richard Herrnstein, just a month before he died, published a revised version of his arguments from the 1970s. This time he teamed up with a conservative political scientist named Charles Murray, whose previous work had criticised the impact of social welfare programmes. Their book, *The*

Bell Curve, proved even more controversial than Jensen's 1969 article.[39]

There was little in *The Bell Curve* that hadn't already been said by the likes of Terman, Jensen or Shockley over previous decades, despite Herrnstein and Murray's claims that they were fearless scientific pioneers confronting issues that others shied away from. Intelligence, they argued, was real and important. Resurrecting Herrnstein's arguments from the early 1970s, they argued that the 20th century had seen the emergence of a cognitive elite, whose wealth and status were rooted in their superior intelligence. At the other end of the social spectrum, an underclass had emerged in which poverty, family breakdown, welfare dependency, poor parenting, criminality and lack of civility was rooted in its members' lack of intelligence. Attempts to raise the intelligence of this group, they claimed, had met with very limited success. Egalitarian social programmes were largely doomed to failure. The exception were gifted education programmes, which were the only government policies with a hope of making a difference.

The book also addressed the genetics of intelligence. Like Shockley and others, Herrnstein and Murray argued that the United States was facing a dysgenic threat – that average intelligence levels were declining. They argued that genetic factors played a role, alongside environmental ones, in intelligence differences. Most controversially of all, they resurrected the idea that socioeconomic inequalities between races were affected by unequal intelligence levels that were, at least in part, genetically determined. It was this claim that catapulted the book to notoriety.

The Bell Curve was a publishing sensation. It sold 400,000 copies in the first two months.[40] It was discussed on prime-time breakfast TV shows in the United States. It provoked a war of words between competing groups of psychologists in the media, laying out for the public different interpretations of what 'mainstream' science believed about the nature of intelligence. And it provoked a huge political reaction. Liberals denounced it as a racist throwback to the biological determinism of the past. Many conservatives liked it for the same reasons their counterparts had embraced the arguments of Jensen, Eysenck and Herrnstein in the late 1960s: because it challenged the liberal, egalitarian status quo which they thought had become dominant in modern American society.

The book, and the controversy surrounding it, also had an impact on American education policies. Supporters of gifted education used the renewed public debates about intelligence and social policy to call for greater funding. Opponents of *The Bell Curve*'s genetic determinism embraced the argument that infant brains were plastic and that intelligence could be boosted through early years education. Under the Clinton administration, Hillary Clinton led a drive for early years education that was founded on neuroscientific arguments about the possibility of raising infant IQ. Liberal enthusiasm for these arguments was a direct response to the storm stirred up by *The Bell Curve*.[41]

But these critics never succeeded in putting the *Bell Curve* genie back in the bottle. As we saw in the very first chapter, the book's ideas about cognitive elites, intelligence and genetics embedded themselves in the public mind and popular culture in the early 21st century.

Rethinking intelligence

As a result of these various scandals, from the 1960s onwards intelligence science and IQ increasingly came to be seen as 'problematic', even while they retained a hold over popular culture. One response to these problems was the complete rejection of intelligence as a basis for managing society. This was evident in the egalitarian turn in education systems, such as the creation of comprehensive schools in England and Wales. Another, embraced by much of the psychological community, was to develop alternatives to rigid ideas about intelligence.

These attempts to rethink intelligence had their roots in the 1970s. It was then that psychologists like Daniel Kahneman and Amos Tversky began to question the notion of humans as rational decision-makers.[42] Instead, they showed that our judgements – even ones involving things like risk and money which we think we approach in considered ways – are shaped by our intuitions, biases and heuristic 'rules of thumb'. Sometimes these kinds of non-rational thought processes are helpful, but often they lead us astray.

Building on these kinds of insights, new psychological theories began to emerge from the 1980s which challenged ideas about the nature and significance of human intelligence. Concepts like multiple intelligences, emotional intelligence, creativity and 'grit' came to be identified as the factors truly underpinning success. They represented alternatives to the old ideas about intelligence which had been discredited by the controversies over race and IQ. More radically, the neurodiversity movement sought to jettison the idea that a single set of psychological characteristics should be regarded as essential to life success.

The era's first big challenge to the hegemony of IQ was the idea of 'multiple intelligences' (MI). The theory of MI was the brainchild of a man named Harold Gardner, a Harvard psychologist with an interest in the psychology of arts and education. In his 1983 book *Frames of Mind*, Gardner argued that there were in fact *seven* distinct human intelligences.[43] Two of those – linguistic intelligence and logical-mathematical intelligence – corresponded fairly closely to the things measured by IQ tests. But Gardner argued that they were combined, in all human beings, with musical intelligence (sensitivity to things like pitch and rhythm), spatial intelligence (visual perception and the ability to create and manipulate mental images), bodily-kinaesthetic intelligence (the ability to control one's body or manipulate objects, as in activities like sport or mime) and two forms of personal intelligence: the ability to understand others and the ability to understand ourselves. In later work he also suggested the possible existence of naturalist intelligence (the ability to understand organisms and the natural world) and existential intelligence (relating to spirituality and 'big questions').

The origins of the theory were entangled with the histories of gifted education and intelligence augmentation that we saw in previous chapters. Gardner carried out his research as part of Harvard's 'Project on Human Potential', which was funded by the Bernard van Leer Foundation, a Dutch NGO that supported work in the fields of education and childcare. The foundation had also been one of the main funders of Don Calogero's Villaggio del Superdotato in Sicily. Gardner had a keen interest in gifted education. The human potential project and another of Gardner's Harvard teams, called 'Project Zero', were also closely involved with the Venezuelan Ministry for the Development of Intelligence. Indeed, Gardner stated that the most important aim behind developing the theory of multiple intelligences was to provide guidance for international policy-makers thinking about ways to raise human intellect.[44] His book discussed some of Luis Alberto Machado's ideas and the policies he was implementing in Venezuela, although it approached them with a degree of scepticism.

Gardner didn't entirely reject mainstream intelligence science, but he framed his theory as a direct response to the ways in which IQ had come to be understood and used in modern society. In contrast to people like Herrnstein and Jensen, who felt that the notion of intelligence and intelligence testing had become taboo in post-war America,

Gardner saw them as both ubiquitous and highly damaging. Indeed, he began his book with the story of a young girl being given an IQ score on the basis of an hour of written tests. This score indelibly influenced her life by shaping the way her teachers thought about her and the educational and professional opportunities which became available to her. Stories like this, he argued, played out thousands of times every week around the world.

Gardner believed that these tests were deeply inadequate, both because of the impact they had on individuals' lives and the fact that they were incapable of measuring the full range of intelligent human behaviours. The problem lay in our ingrained views about intelligence. 'Only if we expand and reformulate our view of what counts as human intellect,' he argued, 'will we be able to devise more appropriate ways of assessing it and more effective ways of educating it.'[45]

In some ways, the theory of multiple intelligence was a return to much older traditions of thinking about human mental abilities which we encountered at the start of this book. Those traditions – in which ideas about mental powers were entangled with different notions of wisdom, character, the divine and so on – were largely abandoned by the intelligence science of the early 20th century, which reduced intelligence to a logical-linguistic core. Gardner, who was deeply interested in the history and anthropology of intelligence, was retrieving some of the traditional ways of thinking about intelligence that had been cast aside in the modern era.

The theory of MI made many intelligence scientists very cross. Some of their criticisms were about terminology. It just didn't make sense, they argued, to label things like bodily-kinaesthetic ability as 'intelligence'. It may be a talent or an ability that people value, but it simply wasn't the kind of thing people mean when they talk about intelligence. Their other big objection was empirical. If these seven intelligences were real, they argued, you should be able to test, measure and assess them, subject them to factor analysis, and do all of the things that psychometricians had been doing with intelligence tests since the start of the century. But the psychometric evidence for MI remained elusive. Gardner, unperturbed, responded that he wasn't interested in psychometric approaches, and that his theory was based on solid neurological, evolutionary and anthropological evidence.

Such criticisms did nothing to dent the theory's popularity. Gardner's book, a fairly dense and technical study, was a surprise hit. It was particularly popular among teachers. Schools immediately began to think about how the theory could change the way they helped their students to learn. Little more than a year after the book was published, a group of teachers in Indianapolis had set up a new school organised around the idea of MI. In the years that followed the idea spread around the world, shaping school curricula, business practices and even the design of theme parks. By the early 2000s, for example, over 100 books had been published about MI in China alone, and many schools in the United States and abroad claimed to have redesigned themselves around MI ideas.[46]

The global enthusiasm for MI tells us something about the way intelligence had come to be understood by the 1980s. The idea of intelligence was still extremely powerful. Gardner's theory was popular precisely because, to the annoyance of intelligence scientists, he insisted on labelling his seven categories as 'intelligence' rather than mere talents or skills. But the way people embraced the theory suggests that many shared Gardner's concerns about the limits and dangers of narrow ideas about intelligence, and about the use of intelligence testing. The controversies over race and IQ of the late 1960s and early 1970s had helped to reinforce these popular concerns. Such concerns were then exacerbated by the *Bell Curve* controversy of the mid-1990s. It was this moment that saw the emergence of the second big popular theory that sought to challenge accepted ideas about IQ: emotional intelligence, or 'EQ'.

The idea of EQ was popularised by an American psychologist and science journalist named Daniel Goleman. Goleman and his wife, psychotherapist Tara Bennett-Goleman, had a long-standing interest in meditation and mindfulness. He had then become fascinated by the new research neuroscientists and psychologists were producing on the science of emotions, including people like Howard Gardner. The result was a book called *Emotional Intelligence: Why It Can Matter More Than IQ*.[47]

Published just a year after the *Bell Curve* controversy, the book explicitly challenged the idea that success in life rested on a person's genetically determined IQ. Instead, Goleman argued, the capacities most valued in the modern world were linked to emotions: our self-awareness and ability to manage our own emotions; our understanding of other people and their emotions; and the ways we use this knowledge to build successful relationships. Brainpower alone wasn't enough to succeed in modern society. It needed to be combined with things like self-control, interpersonal skills, zeal and motivation.

And crucially, Goleman believed, EQ could be learned. Teaching social and emotional skills in schools could help to boost the overall supply of EQ across the population, ameliorating the epidemic of anger, loneliness and depression that seemed to be sweeping modern society. Businesses could also thrive by boosting the EQ of their leaders, helping to create effective teams and harmonious workplaces. In contrast to the dystopian determinism of Murray and Herrnstein, which envisaged a permanent low-IQ underclass, EQ was presented as something that everyone could attain.

The idea of EQ proved hugely successful. Goleman's book spent months on the *New York Times* bestseller list and was translated into over 40 languages. 'EQ' became a recognised term in several global languages. The idea of EQ has become a mainstay of modern business schools, self-help books and dating profiles.

At the heart of Goleman's theory of EQ were the ideas of self-control and delayed gratification. You've probably already heard about the 'marshmallow test' of delayed gratification. The test was first carried out by a psychologist called Walter Mischel at Stanford (yes, we're back there again) in the 1960s. In Mischel's test, an adult offered a four-year-old child a choice: either wait until the adult had run an errand, after which they could have two marshmallows; or eat a marshmallow before the adult returned, in which case they'd only be allowed to have one. There are dozens of videos of this experiment online, showing kids desperately staring at the marshmallow they know they shouldn't eat, sometimes covering their eyes, sometimes licking it or smelling it, and often just giving in and eating it.

The test was designed to measure the ability of children to control their impulses and delay gratification in pursuit of a future goal. Mischel's team tracked their subjects for a number of years. The children who could delay gratification at the age of four seemed to grow into adolescents with better social skills, behaviour, verbal fluency and academic performance. A child's performance on the test, they argued, was a better predictor of later SAT scores than their IQ.

Mischel's test is now famous, but it was almost unknown outside of the psychological community until Daniel Goleman put it at the heart of his theory of EQ in the 1990s. The ability to delay gratification, Goleman argued, was at the heart of emotional self-regulation. It was the thing that made EQ a 'master aptitude' or 'meta-ability', determining how well we're able to make use of our other mental strengths.[48]

These ideas, in turn, led in the 2000s to a new generation of studies that promoted things like perseverance, resilience and 'grit' as the psychological key to a successful life.[49] Like the idea of EQ, these theories were couched in the language of positive psychology. They moved away from a narrow focus on cognitive skills to think more broadly about the importance of psychological strengths and the things people could do to cultivate them. No one, according to these theories, was doomed to a life of failure by their low IQ. All that was required to reach the top was self-discipline. And that was something anyone could learn.

As the historian Michael Staub has argued, however, ideas like EQ and 'grit' didn't entirely succeed in shaking off the troublesome politics of IQ and intelligence science.[50] For a start, these theories understood the psychological roots of poverty and crime as purely a product of a person's mind, rather than of the society around them. Psychologists in the immediate post-war era had tended to understand an individual's psychology as the product of their socioeconomic environment. For example, such psychologists argued, working-class and middle-class communities sometimes valued different traits. They therefore raised their children differently, encouraging the traits they valued most highly. Delayed gratification was often understood as a distinctly middle-class value, and so was more likely to be cultivated in middle-class children. Some psychologists even argued that delaying gratification should be viewed negatively, the product of overly strict parenting that was at the root of middle-class neurosis. But from the 1960s, when Mischel was carrying out his delayed-gratification experiments, these socioeconomic explanations had begun to fall out of fashion. Mischel and his followers saw the test as a measure of an innate ability, not one shaped by the environment children were brought up in. They also assumed that the ability to delay gratification was an inherently positive one.

These different ways of thinking about human characteristics – as either innate or shaped by our environments – could lead to vastly different interpretations of the same phenomena. It would have been possible, for example, to interpret the results of the marshmallow test very differently from the way Mischel did, as a reflection of children's environment rather than their nature. In this interpretation, different behaviours could have been understood as the result of varying levels of trust in authority figures. If you grew up in an environment where parents, teachers or other adults regularly didn't treat you fairly or stick to their word, it would make sense to eat the one marshmallow now rather than relying on the tester to actually bring the two marshmallows they had promised. If we think about the experiment through the prism of race, we could also see it as a test of how much Black children felt they could trust White authority figures. But the testers, and those like Goleman who later promoted their findings, didn't see it this way. They interpreted the tests as a reflection of qualities of impulse control internal to each individual child.

Impulse control was increasingly being seen as the root of all social evils. Criminologists pointed to it as the cause of delinquent behaviour, rooted in permissive and morally lax home environments. People like Herrnstein and Murray argued that self-discipline was one of the characteristics helping the cognitive elite to flourish, and that a lack of self-discipline was at the root of social problems such as 'illegitimate' children. Often these arguments were racialised. Following a long tradition of attributing Black social problems to moral failings, now Black poverty was explained by the inability to delay gratification.

EQ was meant to be a cuddlier alternative to IQ, a way to rescue ideas about mental strengths and abilities in the wake of the race and IQ controversies. But impulse control, the core of EQ, was tied up with a set of assumptions about class and race. And just as with IQ, it explained social hierarchies and inequalities as a product of the contents of people's minds. Proponents of EQ theory could ignore inequalities of class, gender or race. All anyone needed to succeed, they urged, was a bit more grit.

Similar criticisms could be made of MI. The theory is, after all, one still rooted in the assessment of intelligence. As some critics have pointed out, if you don't like the idea of people being negatively judged for their IQ, it doesn't necessarily seem any better to offer a system where people can be told they're stupid in seven different ways. Although it wasn't Gardner's intention to replace one hierarchical framework or ranking method with another, that's often how other people have used his theory.

MI and EQ are far from the only reinterpretations of intelligence that have become popular since the 1980s. Traits like *creativity* have been hailed as the true secrets of success in the modern world. The American commentator Thomas Friedman has argued that 'CQ' – *curiosity quotient* – is more important than IQ in the digital age. Neuroscientists, psychologists and management gurus have challenged our idea that intelligence belongs to individuals by exploring the concept of *collective intelligence*. And over recent years there has been increasing interest in the forms of intelligence evident in the natural world – from octopuses to forests – and the ways they could help us to think differently about human intelligence.

But arguably the most radical challenge to traditional views about human minds and their abilities came from the neurodiversity movement. The idea of neurodiversity encompasses those with a range of conditions, including ADHD and dyslexia. But it emerged primarily from the autistic community.

Autism was first identified in 1943 by two men working simultaneously in different parts of the world.[51] One was Leo Kanner, an Austrian-American psychologist working at Johns Hopkins in Baltimore. He identified a group of children sharing similar characteristics who had typically been given the catch-all diagnosis of 'childhood schizophrenia'. These children were characterised by their 'extreme aloneness', failing to respond to things coming to them from the outside world, anxious when interacting with people, and obsessed with sameness, repetition and patterns. Kanner described these as 'autistic disturbances', suggesting they formed the basis of a distinct and serious condition affecting only a small proportion of the population.[52]

At almost exactly the same time, a paediatrician in Vienna named Hans Asperger was writing his thesis on a group of children with a similar range of conditions. Like Kanner, he reached for the labels of autism and 'autistic psychopathy' to describe children who combined extreme social awkwardness with a fascination for rules, schedules and patterns. In contrast to Kanner, however, Asperger described a more highly functioning group of children. Autistic characteristics, he suggested, were shared by a wide range and variety of people, from the deeply disabled to the profoundly gifted to the slightly eccentric. They were likely, he believed, to be present across a large proportion of the population.

Initially it was Kanner's interpretation that won out. Autism diagnoses tended to be given to a small number of children with severe developmental problems. These problems were often blamed on the emotional failings of parents, and particularly on so-called 'refrigerator mothers'. Autism was regarded as an incurable condition, with many sufferers denied access to proper education and condemned to a lifetime of institutional care. Asperger's broader understanding of autistic characteristics had been lost.

But things began to change in the 1980s. Lorna Wing, a British psychologist working at the Institute of Psychiatry in South London, was herself the mother of an autistic daughter. She helped revolutionise the understanding of autism by resurrecting some of the earlier ideas of Hans Asperger, arguing that autism was a *spectrum* rather than a neatly defined condition. It was defined, she argued, by impairments in social interaction, social communication and social imagination. And she showed that these impairments were much more widely shared across the population than was previously believed. Wing coined the term 'Asperger's syndrome' to describe this range of impairments, a diagnosis which was later folded into the current definition of autism spectrum disorder (ASD).

These new ways of thinking about autism began to change public perceptions. Films like *Rain Man*, in which Dustin Hoffman portrayed an autistic character, helped to popularise the idea that those with autism could function within modern society, and even make valuable contributions to it. Lorna Wing's argument that autistic characteristics were widely spread across the 'normal' population, and indeed that these characteristics might be a key part of scientific and artistic achievement, helped to lay the groundwork for a new generation of activists who saw these characteristics as something to be celebrated.

From the 1990s an autistic community began to emerge in the new online world, communicating through early digital message boards and mailing lists. These digital communities developed new ways of talking and thinking about autistic people and their relationship to modern society. They began to label the non-autistic community as 'neurotypical', parodying attitudes to autism by analysing the delusions and obsessions of those with 'neurotypical syndrome'. Taking inspiration from communities like the Deaf rights movement, they began to challenge the idea of autism as a disability.

This new generation of activists pushed back against the idea that autistic children had their true personalities hidden or trapped in an autistic shell. They challenged those who encouraged the parents of autistic children to grieve for the idealised 'normal' child they had lost. And they began to organise events and spaces where autistic people could come together in real life, in an atmosphere free of the external stimuli and negative stereotypes they faced in neurotypical environments.

In the late 1990s, an Australian researcher named Judy Singer became interested in these online communities, recognising in their discussions experiences from her own life and from those of both her mother and her daughter. She identified these communities as representatives of a new movement. Drawing parallels with the Black Power, feminist and gay liberation movements that had emerged over previous decades, she coined the term *neurodiversity* to describe the struggle for the rights of the neurologically different.

Neurodiversity activists didn't just argue that autism was *not* a disability. They held that in the modern information age, when everyday life was being transformed by computers and digital technologies, neurodivergent minds and ways of thinking were becoming ever more valuable. The success of Silicon Valley seemed to herald the 'rise of the geeks', a community driving economic growth and technological progress where neurodivergent traits were celebrated and rewarded rather than being suppressed. 'Neurodiversity,' a supportive journalist wrote, 'may be every bit as crucial for the human race as biodiversity.'[53]

This was a genuinely radical set of ideas, one that challenged simplistic models of how the mind worked and what mental traits society should value. Society, the neurodiversity movement argued, should celebrate everyone no matter how their brains work or what kind of abilities they can develop. In that sense, it stood in stark contrast to the simplistic, hierarchical idea of intelligence which had emerged at the start of the 20th century. Perhaps neurodiversity offered a better route out of the intelligence cul-de-sac than ideas like MI or EQ?

But the position of intelligence within ideas about neurodiversity was never straightforward. The relationship between autism and intelligence has long been contested. Some, but by no means all, autistic people also suffer from some form of intellectual disability. Historically, children we would now identify as autistic were labelled as idiots or imbeciles, and assumed to be essentially uneducable. But since the 19th century there had also been lots of interest in savants – people with autism-like characteristics who were capable of extraordinary feats of memory, calculation or other mental skills.

The pioneers of the study and treatment of autism often emphasised the intelligence of autistic people. Although the children described by Leon Kanner displayed relatively severe symptoms, he noted their wide vocabulary, excellent memory and ability with patterns, arguing that they indicated 'good intelligence in the sense in which this word is commonly used'.[54] Asperger emphasised that those with autism could span the full range of intelligence. He also identified a particular 'autistic intelligence' among the children he observed – spontaneous, original and experience-based, but not easily captured by mainstream intelligence tests or particularly suitable to traditional methods of education.[55] Both men, reflecting the 19th-century practice of linking character and ability to facial features, commented on the 'intelligent physiognomies' of the children.[56] Since then, public portrayals of autistic people have also often emphasised their intelligence. Raymond Babbitt, Dustin Hoffman's character in *Rain Man*, was portrayed as a savant with incredible mathematical skills, winning thousands of dollars by counting cards in a Las Vegas casino.

But critics of the neurodiversity movement have argued that this emphasis on the high intelligence of autistic people has proved harmful. In a movement almost by definition led by those with high-functioning autism, there has been a natural tendency to celebrate overlooked abilities of the autistic community. But in doing so, critics allege, it has risked undermining research, treatment and recognition for the challenges of those with more severe forms of autism, particularly those with nonverbal autism or with severe intellectual disabilities.[57] Neurodiversity seemed to offer a new way of celebrating human mental abilities. But it couldn't entirely break free from the hierarchical ideas about intelligence that had suffused society over previous centuries.

What should we make of all these movements which have sought, since the 1970s and 1980s, to rethink intelligence and the powers of the human mind?

They've certainly changed the way we talk about minds and the things we value about them. The language of multiple intelligences, emotional intelligence and neurodiversity are now firmly embedded in our culture, from social media to corporate training programmes. They've helped to expand the ways we think about intelligence, recovering some of the complexity and diversity that was gradually lost before the 19th century and then completely buried by the intelligence science of the early 20th century.

Those early 20th-century ideas about intelligence don't have quite the same power over our society as they used to. Intelligence testing plays less of a role in the education systems of many countries than it did in the past. And the recurring controversies over race and intelligence science have done much to damage the popular appeal of IQ.

But as we saw in the very first chapter, this certainly doesn't mean that the idea of intelligence has been discredited. In fact, the new theories rethinking intelligence from the 1980s and the *Bell Curve* controversy of the 1990s coincided with a revived faith in the importance of intelligence in the modern world. The revival of older ideas about cognitive elites in the past two decades is testament to the power that the idea of intelligence continues to hold over us.

10

Why the History of Intelligence Matters

IF YOU'VE READ THIS FAR, I hope I've convinced you that intelligence is not only a mental power rooted in the structure of our brains, it's also an *idea*, and an idea that has a history.

But what do we gain from knowing this? Why does it matter that intelligence has a history?

The way we think about ourselves and the world around us is built upon historical foundations. We like to think that all of our beliefs are the result of a process of careful reasoning from first principles. Reader, they are not. The things we hold to be self-evident are usually the result of a steady accretion of centuries' worth of ideas, culture and popular folklore which have trickled down to us in ways we rarely stop to think about and can never fully understand. As John Maynard Keynes said in relation to economic ideas, '[p]ractical men, who believe themselves to be quite exempt from any intellectual influences, are usually the slaves of some defunct economist. Madmen in authority, who hear voices in the air, are distilling their frenzy from some academic scribbler of a few years back.'[1]

Exactly the same can be said about intelligence. Whenever someone is described as 'smart', or someone else feels their intelligence is being judged because of the job they do or their performance in an exam, or a figure like Elon Musk or Tony Blair says something outlandish about AI, we can see the hand of history reaching into the present. Most of our assumptions about what intelligence is and why it matters have been shaped by some defunct intelligence scientist, some self-help scribbler, some quiz show, some half-remembered comment from a teacher. In short, by all of the history we've been exploring in this book, and lots more besides.

Understanding this history matters because it can help us reflect on our assumptions and beliefs, and on where they might have come from. In doing so it can help us to think more clearly about intelligence in the present

Take AI, for example. I'm more than a little wary of saying anything about the future of AI. Historians are always reluctant to talk about the future because they know that the historical change is a messy and complicated business, that what *has* happened is not a good guide to what *will* happen. In the case of AI, the fundamental science and the technologies built upon it seem to be developing so quickly that by the time you are reading this book, things are likely to look very different from the way they do at the time I'm writing it. Nevertheless, the history of human intelligence clearly has something to offer to our current debates about AI.

Over recent years two big sets of ideas have dominated public discussion on the topic. One set of ideas sees AI as transformative in a positive way, a technology that will revolutionise the economy, science and everyday life. It will cure diseases, provide solutions to mitigate climate change and usher in a new era of general prosperity. The other set of ideas sees AI as transformative in a negative way. It will take our jobs, quickly develop out of our control and possibly end up destroying the human race.

Both of these views are rooted in the history of ideas about human intelligence. They rest on the idea that current technologies will soon develop into artificial general intelligence (AGI) – replicating human levels of intelligence across all fields of mental activity – before accelerating off into some form of superintelligence. But as we know from the early chapters of this book, the notion of 'general intelligence' – which sits at the heart of the concept of AGI – is the product of a particular moment in early 20th-century intelligence science, and one which has always been deeply contested. The frequency with which AI promotors use the language of IQ to sell the technology reveals how in hock they remain to the history of ideas we've been examining here.

These visions of AI are also entangled with the way we understand the historical link between intelligence and progress. This has been one of the defining ways of thinking about intelligence since the late 19th century – that intelligence is the solution to the world's problems, and that society would be transformed if only intelligence could be improved or augmented or harnessed in some way. We've seen these ideas time and again over the course of this book, from the invention of intelligence tests and eugenics to gifted education and the development of the personal computer. Stephen Cave, one of the few people who has explored the relationship between the history of intelligence and AI, describes these types of ideas as the 'fetishization of intelligence'.[2] The belief that AI will be transformative – whether in a positive or negative way – is rooted in this historical association between intelligence and progress.

One thing we might want to reflect on if we're thinking about the transformative power of AI is that many of the projects we've discussed in the book were failures. That's true of the projects that tried to turbocharge progress by boosting human intelligence for all, like Venezuela's Ministry for the Development of Intelligence. And it's true of projects that put their faith in the power of the most intelligent subset of the human population, like Mensa, or the Villaggio del Superdotato, or the Nobel Prize winners' sperm bank. There doesn't seem to be any clear historical evidence that projects that explicitly aimed to harness or improve human intelligence have succeeded in driving social or economic progress. These failures don't provide hard evidence that AI isn't going to transform the world. But they should give us reason to pause before we accept claims that there is some kind of automatic link between AI and progress.

To take a recent example of such claims, Tony Blair wrote in his book *On Leadership* that the capabilities of AI resemble the human brain, but with the added ability to keep improving its intellect. '[A]t present,' Blair writes, 'it [AI] resembles a person with an average IQ of about 100 . . . but as the reinforcement loop gathers momentum AI will soon have an IQ of 150 – that's very smart. In time, it will have one of 200. Then we're in a new world.'[3] Now, we should perhaps be wary of accepting tech advice from the man who claimed to have never even sent an email during his time as prime minister. And we might also raise an eyebrow at the idea that we can learn anything from giving AI an IQ test. But the history we've been discussing over the course of this book should also make us question the easy assumption that ever-smarter AI will automatically usher in a new world.

Progress is one of the big ideas that we've taken from the history of intelligence and dragged into our current debates about AI. Inequality is the other, and it's here that the most important lessons for the way we think about AI might well lie. We've seen throughout this book how the idea of intelligence has been used to justify inequality, to build new hierarchies and defend the status of elites. Our modern idea of intelligence is impossible to disentangle from the history of ideas about the dominance of human beings over the natural world, of White Europeans over other races and regions, of men over women, and of rich over poor.

When we think about the future of AI, then, we're hard-wired to think about it in terms of inequality and subjugation. We can see this in the warnings about the existential risk posed by AI. These warnings draw on our experience of intelligence being used to justify domination and extermination and project it into the future, this time with the human race the one to be subjugated. But we can also see it in the more mundane warning about mass job losses. The rhetoric of recent decades has celebrated 'knowledge workers', offering lavish rewards to the highly skilled in sectors like finance and consulting, while average wages for everyone else have stagnated. Now those same knowledge workers fear it's their turn to have their skills automated. For those who believed that their wealth and status depended on their high intelligence, it's hard not to assume that the arrival of more intelligent machines will undermine that status. We struggle to envisage a world in which humans retain their intrinsic value if they're no longer the most intelligent beings on the planet.

At the same time, though, the history of intelligence shows us that people don't like being told that they're stupid. The 1969 Haringey protests are an example of people successfully challenging discrimination and mistreatment based on spurious arguments about intelligence. If, over the coming years, people begin to lose their jobs to machines en masse while being told that the superiority of those machines makes their own intelligence redundant, it's likely to provoke a lot of resistance. What form this resistance will take is impossible to know. But it's easy to imagine the rise of a kind of Neo-Luddism, a resistance to AI technologies being used to undermine people's living standards and sense of worth. Resistance to AI could emerge as a major social and political force, especially if people feel they're being told they no longer have value because they're not smart enough.

And maybe we should take inspiration from this. Over the past century, we've allowed ideas about intelligence to be used to reinforce and justify inequality. But that doesn't mean we have to continue to do so. There's no objective reason why knowledge work should be valued more than other work. We had a glimpse of the alternative during the Covid lockdowns and the celebration of essential workers, but we forgot it very quickly afterwards. There's no reason to believe that access to lifelong learning and advanced education should depend on a person's performance in a set of exams at the age of 16 or 18. We could return to open admissions to university, or admission by lottery. No one should be made to feel they are less valuable to society because they're not good at the kind of formal study required in school, or because they don't do well in a small number of exams during a couple of years of their childhood. And there's no objective reason why graduates should earn so much more than non-graduates.

Ultimately there's no reason why, in the 21st century, we should try to isolate intelligence and claim it's somehow worth more than other human virtues. Being intelligent is not more important than being wise, or thoughtful, or kind, or patient. Being good with numbers or logic or language isn't inherently better than being good with people. Maybe that's something that AI will actually help us to appreciate better.

This is not to argue that we need to completely abandon the idea of intelligence. We can still recognise and celebrate intelligent behaviour when we see it. Scientists can still continue to try to uncover its secrets – whether in humans, animals or machines. We should continue to value the way education expands our mental capacities and horizons. Human intelligence is an incredible thing, and we don't need to pretend it isn't.

But what we can stop doing is taking the idea of intelligence and building grand theories of inequality on top of it, trying to pretend that differences in intelligence should structure our society. We can resist those who try to use the idea of intelligence to justify hierarchy and inequality, to argue that some human beings are worth more than others. The history of intelligence offers a foundation on which to base that resistance.

As I said in the introduction, understanding this history will not make us any smarter, but it might help us to be a bit wiser.

Acknowledgements

I've discussed this book with many friends, family and colleagues over the last few years, and almost everything I've written here has been shaped by those conversations. I fear if I try and list everyone I will inevitably leave someone out, so I will instead thank you all collectively. Thank you!

I would, however, particularly like to thank my colleagues Caitjan Gainty, Sundeep Lidher, Chris Manias, Hannah Murphy and Chris Parkes, who kindly agreed to read parts of the manuscript and to offer their hugely insightful advice and feedback.

Thank you to Adam Gauntlett for his support at the start of the process, and to Fritha Saunders, Justine Taylor and everyone else at Footnote and Bonnier who have contributed to the book's development since then.

Above all, thank you to Katie for reading and commenting on every single chapter of the book, and for making it immeasurably better than it otherwise would have been.

Joyce and Albert are the best.

Notes

Introduction
1. OECD, *Education at a Glance 2022: OECD Indicators* (Paris: OECD Publishing, 2022), https://doi.org/10.1787/3197152b-en
2. Stephen J. Gould, *The Mismeasure of Man* (New York: Norton, 1981). Gould's book, like most of the scholarship on the history of intelligence, concentrates on intelligence science in general and IQ in particular. This book seeks to decentre and contextualise intelligence science, situating it within the broader history of 20th-century politics, technology and culture, with a particular focus on the history of inequality. In doing so it builds on works from different disciplinary perspectives, such as the philosopher Catherine Malabou's *Morphing Intelligence: From IQ Measurement to Artificial Brain* (New York: Columbia University Press, 2019), which explores the intersections between intelligence science, neuroscience and technology. It also extends the work of historians who have taken a similar approach to earlier periods, most notably John Carson, *The Measure of Merit: Talents, Intelligence, and Inequality in the French and American Republics, 1750–1940* (Princeton, NJ: Princeton University Press, 2007). And, for the first time, it brings together distinct topics and fields of scholarship within the broader history of intelligence, such as Stephen Cave and Kanta Dihal's work on AI narratives and imaginaries, Michael E. Staub's scholarship on race and intelligence, and a substantial scholarship on the history of education and meritocracy (see relevant notes for further details).

Chapter 1: The Cognitive Elite

1. Bethany McLean and Peter Elkind, *The Smartest Guys in the Room: The Amazing Rise and Scandalous Fall of Enron* (London: Penguin, 2004).
2. 'The rise and rise of the cognitive elite', *Economist* (22 January 2011).
3. Daniel Bell, *The Coming of Post-Industrial Society: A Venture in Social Forecasting* (New York: Basic Books, 1973).
4. Robert Reich, *The Work of Nations: Preparing Ourselves for 21st Century Capitalism* (New York: Vintage Books, 1992).
5. 'The e-generation is with us', *Guardian* (7 March 2000).
6. Michael J. Sandel, *The Tyranny of Merit: What's Become of the Common Good?* (London: Allen Lane, 2020), 20.
7. Adrian Wooldridge, *Measuring the Mind: Education and Psychology in England, c.1860–c.1990* (Cambridge: Cambridge University Press, 1994), 253.
8. Ben Young and Robert Hazell, *Putting the Goats Amongst the Wolves: Appointing Ministers From Outside Parliament* (London: The Constitution Unit, UCL, 2011).
9. Christopher D. McKenna, *The World's Newest Profession: Management Consulting in the Twentieth Century* (Cambridge: Cambridge University Press, 2006).
10. *The Mensa Correspondence*, 48 (February 1963), 6.
11. Duff McDonald, *The Firm: The Inside Story of McKinsey: The World's Most Controversial Management Consultancy* (London: Oneworld, 2014), 42.
12. Ibid., 45–6.
13. Daniel Markovits, *The Meritocracy Trap* (London: Allen Lane, 2019), 163.
14. Ibid., 295.
15. 'War on Wall Street', *Economist* (26 October 1996), 116–20.
16. 'Mark Zuckerberg: Hiring the Right People', (May 2011), https://youtu.be/JPHVeQ7-ynA?si=oQGTrkpv0gkpT4Q1 (accessed 8 January 2025).

17. Walter Isaacson, *Steve Jobs* (London: Abacus, 2011), 334.
18. 'We fund smart founders, irrespective of what they work on', (February 2024), https://youtu.be/sjRtaX35TJM?si=Y4FHI56O HIHBhG0Z (accessed 8 January 2025).
19. Founders Fund manifesto, https://foundersfund.com/2018/01/manifesto (accessed 15 February 2025).
20. Paul Allen, 'Microsoft's Odd Couple', *Vanity Fair* (May 2011), 130.
21. Isaacson, *Steve Jobs*, 158.
22. 'Steve Jobs – Speech to the Academy of Achievement June 1982', https://youtu.be/ymbD_a-G1IQ?si=WsWt2B7uQf FGDaMh (accessed 8 January 2025).
23. Quoted in 'Silicon Valley billionaires remain in thrall to the cult of the geek', *Financial Times* (28 November 2024).
24. Elizabeth G. Chambers, Mark Foulon, Helen Handfield-Jones, Steven M. Hankin and Edward G. Michaels, 'The War for Talent', *The McKinsey Quarterly*, 3 (1998), 45–57.
25. Ibid., 53.
26. Malcolm Gladwell, 'The Talent Myth', *New Yorker* (22 July 2002).
27. Phillip Brown, Hugh Lauder and David Ashton, *The Global Auction: The Broken Promises of Education, Jobs, and Incomes* (Oxford: Oxford University Press, 2011).
28. Quoted in Sandel, *The Tyranny of Merit*, 86.
29. *UNESCO Higher Education Global Data Report* (UNESCO, 2020).
30. 'Tony Blair's full speech' (28 September 1999), https://www.theguardian.com/politics/1999/sep/28/labourconference.labour14 (accessed 28 January 2025).
31. House of Commons Library, *Higher Education Student Numbers* (7 January 2025), https://researchbriefings.files.parliament.uk/documents/CBP-7857/CBP-7857.pdf (accessed 28 January 2025).
32. Martin Carnoy *et al.*, *University Expansion in a Changing Global Economy: Triumph of the BRICs?* (Stanford, CA: Stanford University Press, 2013).

33. Sandel, *The Tyranny of Merit*, 61 & 176.
34. Ben Wildavsky, *The Great Brain Race: How Global Universities are Reshaping the World* (Princeton, NJ: Princeton University Press, 2010).
35. Brown *et al.*, *The Global Auction*, 94.
36. Sandel, *The Tyranny of Merit*, 90.
37. Quoted in Rakesh Khurana, *From Higher Aims to Hired Hands: The Social Transformation of American Business Schools and the Unfulfilled Promise of Management As a Profession* (Princeton: Princeton University Press, 2007).
38. Björn Hammarfelt, Sarah de Rijcke and Paul Wouters, 'From Eminent Men to Excellent Universities: University Rankings as Calculative Devices', *Minerva*, 55 (2017), 391–411.
39. Stefan Wilbers and Jelena Brankovic, 'The emergence of university rankings: a historical-sociological account', *Higher Education*, 86 (2023), 733–50.
40. Wildavsky, *The Great Brain Race*, 104–22.
41. Markovits, *The Meritocracy Trap*.
42. 'Remarks by the President at College Opportunity Summit', 4 December 2014, https://obamawhitehouse.archives.gov/photos-and-video/video/2013/07/08/president-obama-speaks-new-management-agenda (accessed 6 February 2025). Also quoted in Sandel, *The Tyranny of Merit*.
43. Markovits, *The Meritocracy Trap*, 193.
44. Joe Rogan Experience #1169 (7 September 2018), https://www.youtube.com/live/ycPr5-27vSI?feature=shared (accessed 15 February 2025).
45. Walter Isaacson, *Elon Musk* (London: Simon & Schuster, 2023).
46. Erik Baker, 'The Rise of Entrepreneurial Management Theory in the United States', *Modern Intellectual History*, 20 (2023), 195–219.
47. Joseph Schumpeter, *Capitalism, Socialism, and Democracy* (London: Routledge, 2010).

48. Dieter Plehwe, 'Schumpeter Revival? How Neoliberals Revised the Image of the Entrepreneur', in Dieter Plehwe, Quinn Slobodian and Philip Mirowski (eds), *Nine Lives of Neoliberalism* (London: Verso, 2020), 120–42.
49. Quinn Slobodian, 'The Unequal Mind: How Charles Murray and Neoliberal Think Tanks Revived IQ', *Capitalism*, 4:1 (2023), 73–108; Quinn Slobodian, *Hayek's Bastards: Race, Gold, IQ, and the Capitalism of the Far Right* (New York: Zone Books, 2025).
50. Charles Murray, *Human Diversity: The Biology of Gender, Race, and Class* (New York: Twelve, 2020).
51. Jordan Peterson (18 February 2018), https://www.youtube.com/live/8sSe6FSrylc?si=UCDbxdCBOZTBAZ-q (accessed 19 Feb 2025).
52. Isaacson, *Elon Musk*, 418.
53. Du Rove's Channel, *Telegram* (29 July 2024).
54. 'Who broke the sperm bank?', *Financial Times* (4 March 2025).
55. Most records about the work of the FX Future Fund disappeared with the fund itself. Luckily, some information posted by fund staff on EA forums remains available, as does, in this case, snapshots of its website captured by the WayBack Machine: https://web.archive.org/web/20221114025921/https://ftx.tghp.co.uk/area-of-interest/economic-growth/ (accessed 19 February 2025).
56. Future Fund June 2022 Update, https://forum.effectivealtruism.org/posts/paMYXYFYbbjpdjgbt/future-fund-june-2022-update (accessed 19 February 2025).
57. 'Ending radical and wasteful government DEI programs and preferencing', 20 January 2025, https://www.whitehouse.gov/presidential-actions/2025/01/ending-radical-and-wasteful-government-dei-programs-and-preferencing/ (accessed 1 March 2025).
58. Alexandr Wang, 'Today we've formalized an important hiring activity', https://www.linkedin.com/posts/alexandrwang_today-weve-formalized-an-important-hiring-activity-7207139343713329152-NKLD/ (accessed 1 March 2025).

Chapter 2: Intelligence Before IQ

1. Marcel Detienne and Jean-Pierre Vernant, *Cunning Intelligence in Greek Culture and Society* (Hassocks: Harvester Press, 1978).
2. *The Republic of Plato*, trans. Benjamin Jowett, 3rd ed. (Oxford: Oxford University Press, 1908). Plato's discussion about philosopher kings and their abilities is largely found in sections V–VII.
3. Benjamin Isaac, *The Invention of Racism in Classical Antiquity* (Princeton, NJ: Princeton University Press, 2004).
4. Sari Nusseibeh, *The Story of Reason in Islam* (Stanford, CA: Stanford University Press, 2017).
5. Henry Corbin, *History of Islamic Philosophy*, trans. Liadain Sherrard (Abingdon: Routledge, 2014 [first published 1964]), 162–4.
6. Benjamin A. Elman, *Civil Examinations and Meritocracy in Late Imperial China* (Cambridge, MA: Harvard University Press, 2013), 318. Most of the details about the examination system in the following paragraph are taken from Elman's study.
7. Alexander Woodside, *Lost Modernities: China, Vietnam, Korea, and the Hazards of World History* (Cambridge, MA: Harvard University Press, 2006).
8. Michael Puett, 'Political Theologies of Justice: Meritocratic Values from a Global Perspective', in Tarun Khanna and Michael Szonyi (eds), *Making Meritocracy: Lessons from China and India, from Antiquity to the Present* (Oxford: Oxford University Press, 2022), 32.
9. Hong Xiao and Chenyang Li, 'China's Meritocratic Examinations and the Ideal of Virtuous Talents', in Daniel Bell and Chenyang Li (eds), *The East Asian Challenge for Democracy: Political Meritocracy in Comparative Perspective* (Cambridge: Cambridge University Press, 2013), 340–62.
10. Sudev Sheth and Lawrence L.C. Zhang, 'Locating Meritocracy in Early Modern Asia', in Khanna and Szonyi (eds), *Making Meritocracy* (Cambridge: Cambridge University Press, 2023), 93.

11. Nicolas Tackett, *The Destruction of the Medieval Chinese Aristocracy* (Cambridge, MA: Harvard University Press, 2014).
12. R.P. Dore, 'Talent and the Social Order in Tokugawa Japan', *Past & Present*, 21:1 (April 1962), 60–72.
13. John White has argued that these Puritan traditions were central to the development of intelligence science from the late 19th century. John White, *Intelligence, Destiny and Education* (Abingdon: Routledge, 2006).
14. Laura C. Ball, 'The Genius in History: Historiographic Explorations', in Dean Keith Simonton (ed.), *The Wiley Handbook of Genius* (Chichester: John Wiley & Sons, 2014), 3–20. Ann Jefferson, *Genius in France: An Idea and its Uses* (Princeton, NJ: Princeton University Press, 2015).
15. Jefferson, *Genius*, 31.
16. Alexander Marr, Raphaële Garrod, José Ramón Marcaida and Richard J. Oosterhoff (eds), *Logodaedalus: Word Histories of Ingenuity in Early Modern Europe* (Pittsburgh, PA: University of Pittsburgh Press, 2018).
17. C.F. Goodey, *A History of Intelligence and 'Intellectual Disability': The Shaping of Psychology in Early Modern Europe* (London: Routledge, 2011), 40–3.
18. Carson, *The Measure of Merit*, 25.
19. *Ibid.*, 11.
20. Goodey, *A History of Intelligence*, 129.
21. Simon Jarrett, *Those They Called Idiots: The Idea of the Disabled Mind from 1700 to the Present Day* (London: Reaktion Books, 2020).
22. *Ibid.*, 106.
23. David N. Livingstone, *Adam's Ancestors: Race, Religion, and the Politics of Human Origins* (Baltimore: Johns Hopkins University Press, 2008).
24. Arthur de Gobineau, *The Inequality of Human Races*, trans. Adrian Collins (London: William Heinemann, 1915), 180–1.

25. Herbert Spencer, 'The Comparative Psychology of Man', *The Journal of the Anthropological Institute of Great Britain and Ireland*, 5 (1876), 301–16.
26. Carson, *The Measure of Merit*, 75; Thomas Fuller Obituary, *Columbian Centinel* (29 December 1790), 123.
27. Olaudah Equiano, *The Interesting Narrative of the Life of Olaudah Equiano, or Gustavus Vassa, the African. Written by Himself. Vol. I* (London: 1789), 43.
28. Frederick Douglass, *The claims of the negro, ethnologically considered: An address before the Literary Societies of Western Reserve College, At Commencement, July 12, 1854* (Rochester: 1854), 8.
29. Anténor Firmin, *The Equality of the Human Races* (Urbana: University of Illinois Press, 2000), 163. Thanks to Edouard Beretti-Cahen for bringing Firmin's arguments to my attention.
30. Goodey, *A History of Intelligence*.
31. Mary Wollstonecraft, *A Vindication of the Rights of Women with Strictures on Political and Moral Subjects* (1792), ch. 2.
32. Lucy Delap, 'The Superwoman: Theories of Gender and Genius in Edwardian Britain', *The Historical Journal*, 47:1 (2004), 101–26.
33. Sara Lyons, *Assessing Intelligence: The Bildungsroman and the Politics of Human Potential in England, 1860–1910* (Edinburgh: Edinburgh University Press, 2023).

Chapter 3: The Birth of Intelligence Science

1. Nicholas Wright Gillham, *A Life of Sir Francis Galton: From African Exploration to the Birth of Eugenics* (Oxford: Oxford University Press, 2001).
2. Francis Galton, *Hereditary Genius: An Inquiry into its Laws and Consequences* (London: Macmillan and Co., 1869), 14.
3. Ibid., 21–6.
4. Ibid., v.

5. Ibid., 339.
6. Ibid., 343.
7. Theta H. Wolf, *Alfred Binet* (Chicago: University of Chicago Press, 1973).
8. Alfred Binet, *L'étude expérimentale de l'inteligence* (Paris: Schleicher Fréres & Co., 1902).
9. Alfred Binet and Théodore Simon, *The Development of Intelligence in Children (The Binet-Simon Scale)*, trans. Elizabeth S. Kite (Baltimore: Williams & Wilkins, 1916), 10f.
10. Binet, *L'étude*, 6.
11. Binet and Simon, *The Development*, 202–3.
12. Ibid., 256.
13. C. Spearman, 'General Intelligence,' Objectively Determined and Measured', *The American Journal of Psychology*, 15:2 (April 1904), 201–92
14. C. Spearman, *The Abilities of Man: Their Nature and Measurement* (London: Macmillan, 1927), 76.
15. Carson, *The Measure of Merit*, 185.
16. Lewis M. Terman, *Condensed Guide for the Stanford Revision of the Binet-Simon Intelligence Tests* (Cambridge, MA: The Riverside Press, 1920).
17. 'Condensed Summary of the Preliminary Report of Gifted Children Investigation', Lewis Madison Terman Papers, SC0038, box 10, folder 28, Stanford University Special Collections (SUSC).
18. Terman papers, box 13, folder 41, SUSC.
19. Philip H. DuBois, *A History of Psychological Testing* (Boston, MA: Allyn and Bacon, 1970), 66.
20. Paul Davis Chapman, *Schools as Sorters: Lewis M. Terman, Applied Psychology, and the Intelligence Testing Movement, 1890–1930* (New York: New York University Press, 1988), 101.
21. Terman Papers, box 20, SUSC.

22. Gerd Gigerenzer, 'The Idea of a Peculiarly Female Intelligence: A Brief History of Bias Masked as Science', in Robert J. Sternberg and David D. Preiss (eds), *Intelligence in Context: The Cultural and Historical Foundations of Human Intelligence* (Cham: Palgrave Macmillan, 2023), 93–120.
23. Stephanie A. Shields, 'The Variability Hypothesis: The History of a Biological Model of Sex Differences in Intelligence', *Signs*, 7:4 (1982), 769–97.

Chapter 4: Mental Engineering

1. Diarist 5390, 19th–21st September 1942, Mass Observation Archive, University of Sussex Special Collections.
2. W.E.B. Du Bois, *Darkwater: Voices From Within the Veil* (New York: Harcourt, Brace and Howe, 1920), 10.
3. Willard B. Gatewood, *Aristocrats of Color: The Black Elite, 1880–1920* (Bloomington, IN: Indiana University Press, 1990), 310.
4. W.E.B. Du Bois, 'The Talented Tenth', in Booker T. Washington (ed.), *The Negro Problem* (J. Pott & Company, 1903).
5. Manning Marable, *W.E.B. Du Bois: Black Radical Democrat* (London: Routledge, 2005), 195–6.
6. Michael Collins, 'W.E.B. Du Bois's Neurological Modernity: I.Q., Afropessimism, Genre', in The Triangle Collective (eds.), *The Palgrave Handbook of Twentieth and Twenty-First Century Literature and Science* (Palgrave Macmillan, 2020), 559–76.
7. Jaap van Ginneken, *Crowds, psychology, and politics, 1871–1899* (Cambridge: Cambridge University Press, 1992), 131–86.
8. Gustave Le Bon, *The Crowd: A Study of The Popular Mind* (London: T. Fisher Unwin, 1920), 15.
9. Ibid., 6.
10. Ibid., 32.
11. Ibid., 19.

12. Ibid., 73–4.
13. José Ortega y Gasset, *The Revolt of the Masses* (New York: Norton & Co., 1932).
14. José Ortega y Gasset, *España Invertebrada: bosquejo de algunos pensamientos históricos, segunda edición, revisada y aumentada* (Madrid: Calpe, 1922), 139.
15. Ibid., 136–7.
16. Paul Valéry, *The Outlook for Intelligence* (Princeton, NJ: Princeton University Press, 1962), 79.
17. John Erskine, *The Moral Obligation to be Intelligent* (Indianapolis: Bobbs-Merrill, 1921).
18. Binet and Simon, *The Development of Intelligence*, 262.
19. Henry Herbert Goddard, *Human Efficiency and Levels of Intelligence* (Princeton, NJ: Princeton University Press, 1920), vii.
20. Davis Chapman, *Schools as Sorters*, 68
21. Clarence S. Yoakum and Robert M. Yerkes (eds), *Army Mental Tests* (New York: Henry Holt, 1920).
22. Ibid., vii.
23. Ibid., 191–9.
24. Philip H. DuBois, *A History of Psychological Testing* (Boston, MA: Allyn and Bacon, 1970), 83–119.
25. Charles S. Myers, *Mind and Work: The Psychological Factors in Industry and Commerce* (London: University of London Press, 1920).
26. NIIP/3/1, National Institute of Industrial Psychology papers, LSE Archives and Special Collections (LSEA).
27. 'Vocational guidance and selection work carried out at the Institute', 23 March 1931, NIIP/10/2, LSEA.
28. 'Hints on Choosing a Career' (undated), NIIP/12/6, LSEA.
29. Goddard, *Human Efficiency*, 60.
30. Nicholas Lemann, *The Big Test: The Secret History of American Meritocracy* (New York: Farrar, Straus and Giroux, 2000).
31. Wooldridge, *Measuring the Mind*, 166–72.

32. Gillian Sutherland, *Ability, Merit and Measurement: Mental Testing and English Education, 1880–1940* (Oxford: Clarendon Press, 1984), 98.
33. Wooldridge, *Measuring the Mind*, 175.
34. Ibid., 184.
35. Sutherland, *Ability*, 148–9.
36. Erik Linstrum, *Ruling Minds: Psychology in the British Empire* (Cambridge, MA: Harvard University Press, 2016), 83–99.
37. Richey to Terman, 30 November 1922, folder 38, box 12, Terman Papers, SUSC.
38. Chloe Campbell, *Race and Empire: Eugenics in Colonial Kenya* (Manchester: Manchester University Press, 2007).
39. W.B. Mumford and C.E. Smith, 'Racial Comparisons and Intelligence Testing', *Journal of the Royal African Society*, 37:146 (1938), 46–57.
40. *Education in Africa: A Study of West, South, and Equatorial Africa by the African Educational Commission, under the Auspices of the Phelps-Stokes Fund and Foreign Mission Societies of North America and Europe* (New York: Phelps-Stokes Fund, 1922).
41. Quoted in Shoko Yamada, *'Dignity of Labour' for African Leaders: The Formation of Education Policy in the British Colonial Office and Achimota School on the Gold Coast* (Mankon, Bamenda: Langaa Research & Publishing, 2018), 92.
42. *Educational Policy in British Tropical Africa. Memorandum Submitted to the Secretary of State by the Advisory Committee on Native Education in the British Tropical African Dependencies* (London: His Majesty's Stationery Office, 1925), 261.
43. Yamada, *Dignity of Labour*, 206.
44. *Educational Policy in British Tropical Africa*, 262.
45. Quoted in Campbell, *Race and Empire*, 135.
46. *African Education: A Study of Educational Policy and Practice in British Tropical Africa* (Oxford: Oxford University Press, 1953), 75–6.

47. Terman, *The Measurement of Intelligence*, 11.
48. E.L. Thorndike, *Human Nature and the Social Order* (New York: Macmillan, 1940), 957.
49. Matthew Connelly, *Fatal Misconception: The Struggle to Control World Population* (Cambridge, MA: Harvard University Press, 2008), ch. 2.
50. Julia Barbara Köhne, 'The Cult of Genius in Germany and Austria at the Dawn of the Twentieth Century', in Joyce E. Chaplin and Darrin M. McMahon, *Genealogies of Genius* (Basingstoke: Palgrave Macmillan, 2016), 115–36.
51. Helen Roche and Lisa Pine, 'The Biopolitics of Education in the Third Reich's "Special Schools" and "Elite Schools"', *The Historical Journal*, 66:2 (2023), 413–34.
52. Hitler's speech before the first Greater German Reichstag, 30 January 1939, https://ghdi.ghi-dc.org/pdf/eng/English35.pdf (accessed 16 April 2025).
53. Lippmann's articles and the responses from Terman are reproduced in N.J. Block and Gerald Dworking, *The IQ Controversy: Critical Readings* (New York: Pantheon Books, 1976), 2–44. See also Tom Arnold-Forster, 'Democracy and Expertise in the Lippmann-Terman Controversy', *Modern Intellectual History*, 16:2 (2019), 561–92.

Chapter 5: Mensa

1. Susanne Schregel, '"The intelligent and the rest": British Mensa and the contested status of high intelligence', *History of the Human Sciences*, 33:5 (2020), 12–36.
2. Victor Serebriakoff, *Mensa: The Society for the Highly Intelligent* (London: Constable, 1985), 23.
3. Serebriakoff, *Mensa*, 31.
4. Ibid., 11.
5. *Mensa* (1957).
6. *The Mensa Correspondence*, 22 (December 1960).

7. 'Character and Beliefs Survey: Interim Report', *The Mensa Proceedings,* 7:1 (September 1958).
8. Ibid.
9. Ibid.
10. 'The Mensa Test as Confidence-Booster', *New York Times* (14 April 1999).
11. *The Mensa Magazine, Old Series*, 1:1 (April 1947), 9.
12. *Mensa* (1959).
13. 'With Top Brains It's Talk That Counts', *Sunday Times* (19 November 1961).
14. *The Mensa Magazine, Old Series*, 1:1 (April 1947), 14.
15. *The Mensa Quarterly*, 2:2 (February 1950), 4.
16. *Mensa Annual Report* (1961).
17. *The Mensa Proceedings*, 7:1 (September 1958).
18. *The Mensa Magazine, Old Series*, 1:1 (April 1947), 9.
19. Serebriakoff, *Mensa*, 124.
20. 'Parents Put Bright Children to the Test', *Sunday Times* (9 April 1989).
21. 'Mensa to Open School for the UK's Brightest', *Sunday Times* (11 November 1990).
22. Victor Serebriakoff, *IQ: A Mensa Analysis and History* (London: Hutchinson, 1966), 15.
23. Ibid., 21.
24. Peter Mandler, *The Crisis of Meritocracy: Britain's Transition to Mass Education Since the Second World War* (Oxford: Oxford University Press, 2020), ch. 4.
25. Serebriakoff, *Mensa*, 182–90.
26. Ibid., 231.
27. *British Mensa Magazine* (June 1972).
28. Box 12, folder 8, Nathaniel Weyl papers (NWP), Hoover Institution Library and Archive, Stanford, California (HILA).
29. Boxes 8 and 12, NWP, HILA.
30. Weyl, 'Envy and Aristocide in Underdeveloped Countries' (1974) and 'Israel and South Africa: Two Beleaguered Elites', *Mankind*

Quarterly, 8:3 (January–March 1973), 158–65, box 34, folders 6 and 10, HILA.
31. *Mensa* (1957).
32. *The Mensa Quarterly*, 1:1 (August 1948).
33. *The Mensa Proceedings*, 4:1 (November 1955).
34. Julian Huxley, *UNESCO: Its Purpose and Philosophy* (Washington DC: Public Affairs Press, 1947).
35. Serebriakoff, *IQ*.
36. 'Experts pour scorn on superdads', *Sunday Times* (2 March 1980).
37. 'I think, therefore I am on the way to extinction', *Sunday Times* (7 March 1983).
38. 'Mensa sperm bank set up to create "superhumans"', *Sunday Times* (11 February 1996).
39. *New York Times*, (14 January 1995).
40. *The Mensa Proceedings,* 7:2 (March 1959).
41. Serebriakoff, *IQ*, 73.
42. Serebriakoff, *Mensa*, 81.
43. Madsen Pirie, *Think Tank: The Story of the Adam Smith Institute* (London: Biteback, 2012), 95.
44. Eamonn Butler and Madsen Pirie, *Test Your IQ* (London: Pan Books, 1983), 17–18.
45. 'Freedom Fighter', *Mensa Magazine* (January 2009).
46. Serebriakoff, *Mensa*, 253–4.
47. Ibid., 267.

Chapter 6: Gifted and Talented
1. 'Forging a human being' (undated), box 49, folder 6, NWP, HILA.
2. 'Report on recruitment and selection of children to be granted admission to the villaggio del superdotato "M. Carollo" at Petralia Soprana', folder 6, box 49, NWP, HILA.
3. Jo Danna, *The Sicilian Project: When Ancient Ways Collide with the Modern World* (Xlibris, 2011), 224.

4. Margaret Mead to Henry Saltzman, 3 July 1968, reproduced in 'Forging a New Society', folder 6, box 49, NWP, HILA.
5. Paul R. Gregory, *Lenin's Brain and Other Tales from the Secret Soviet Archives* (Stanford, CA: Hoover Institution Press, 2008), 25.
6. Marina Bentivoglio, 'Cortical structure and mental skills: Oskar Vogt and the legacy of Lenin's brain', *Brain Research Bulletin*, 47:4 (1998), 291–6.
7. Jochen Richer, 'Pantheon of Brains: The Moscow Brain Research Institute, 1925–1936', *Journal of the History of Neurosciences*, 16:1–2 (2007), 138–49.
8. Leon Trotsky, 'From a report at the Seventh All Russian Party conference of April 5th, 1923', Marxists Internet Archive, https://www.marxists.org/archive/trotsky/1925/lenin/12.htm (accessed 10 July 2024).
9. Sheila Fitzpatrick, *Education and Social Mobility in the Soviet Union, 1921–1934* (Cambridge: Cambridge University Press, 1979). The discussion of Soviet education below is largely based on Fitzpatrick's work.
10. Michael David-Fox, *Revolution of the Mind: Higher Learning Among the Bolsheviks, 1918–1929* (Ithaca: Cornell University Press, 1997).
11. Andy Byford, *Science of the Child in Late Imperial and Early Soviet Russia* (Oxford: Oxford University Press, 2020), 224.
12. John Dunstan, *Paths to Excellence and the Soviet School* (Windsor: NFER, 1978), 27.
13. Irina Sirotkina, 'Cultivating Genius in a Bolshevik Country', in Joyce E. Chaplin and Darrin M. McMahon (eds), *Genealogies of Genius* (Basingstoke: Palgrave Macmillan, 2016), 137–51.
14. Irina Leopoldoff, 'A Psychology for Pedagogy: Intelligence Testing in USSR in the 1920s', *History of Psychology*, 17:3 (2014), 187–205.
15. A.R. Luria, *Cognitive Development: Its Cultural and Social Foundations* (Cambridge, MA; Harvard University Press, 1976), 109.

16. Byford, *Science of the Child*, 222–8.
17. Ibid., 244–53.
18. Quoted in Dunstan, *Paths to Excellence*, 25.
19. Boris Kukushkin, 'The Olympiad Movement in Russia', *International Journal of Educational Research*, 25:6 (1996), 553–62.
20. Dunstan, *Paths to Excellence*, 120.
21. Ibid., 219–39.
22. Izaak Wirszup, 'The First Two International Mathematical Olympiads for Students of Communist Countries', *The American Mathematical Monthly*, 69:2 (February 1962), 150–5.
23. Nura D. Turner, 'A Historical Sketch of the Olympiads, National and International', *The American Mathematical Monthly*, 85:10 (December 1978), 802–7.
24. Ida Jeltova and Elena L. Grigorenko, 'Systematic Approaches to Giftedness: Contributions of Russian Psychology', in Robert J. Sternberg and Janet E. Davidson (eds), *Conceptions of Giftedness* (Cambridge: Cambridge University Press, 1986), 171–86.
25. Susanne Schregel, 'Ein "gefährliches Instrument in den Händen der herrschenden Klasse der bürgerlich-kapitalistischen Gesellschaftsordnung" – Intelligenz(test)kritik und Begabungsdeutungen in der frühen DDR (1949–1963)', in Stefanie Coché and Hedwig Richter (eds), *Legitimierung staatlicher Herrschaft in Demokratien und Diktaturen: Festschrift für Ralph Jessen. Politik und Gesellschaftsgeschicht* (Bonn: Dietz, 2020), 187–207.
26. Albert Ziegler and Heidrun Stöger, 'The German View of Giftedness', in Shane N. Phillipson and Maria McCann (eds), *Conceptions of Giftedness: Sociocultural Perspectives* (Mahwah, NJ: Lawrence Erlbaum Associates, 2007), 65–98.
27. Terman papers, box 10, folders 50 & 51, SUSC.
28. Leslie Margolin, *Goodness Personified: The Emergence of Gifted Children* (New York: Aldine de Gruyter, 1994), 35.

29. Lewis Terman, 'Needed endowment for research on education and psychology', Terman papers, box 10, folder 1, SUSC.
30. Ibid., 10–15.
31. Ibid., 38–9.
32. Ibid., 77.
33. Crane, 'Gifted Children'.
34. Jennifer Crane, 'Britain and Europe's Gifted Children in the Quests for Democracy, Welfare and Productivity, 1970–1990', *Contemporary European History*, 32 (2023), 235–53.
35. Quoted in Jolly, *A History*, 78.
36. Ibid., 92–167.
37. Jennifer L. Jolly, *A History of American Gifted Education* (New York: Routledge, 2018), 70.
38. Margolin, *Goodness Personified*, 29
39. Leta S. Hollingworth, *Gifted Children: Their Nature and Nurture* (New York: Macmillan, 1927), 69.
40. Margolin, *Goodness Personified*, 19–20.
41. Margolin, *Goodness Personified*, 23; Jennifer Crane, 'Gifted Children, Youth Culture, and Popular Individualism in 1970s and 1980s Britain', *The Historical Journal*, 65 (2022), 1,418–41.
42. Ibid., 33–5.
43. The material below draws from James Maguire's magisterial *American Bee: The National Spelling Bee and the Culture of Word Nerds* (Rodale, 2006).
44. Andrée Blouin, *My Country, Africa: Autobiography of the Black Pasionaria* (London: Verso, 2025), 142.

Chapter 7: Building an Artificial Brain

1. 'Science: The Thinking Machine', *Time*, 53 (24 January 1949).
2. 'Checkmate in 390,625', *Daily Mail* (13 December 1948).
3. 'Artificial Brain May Solve World's Economic & Political Problems', *The Sunday Indian Express* (2 January 1949).

4. Stephen Cave, Kanta Dihal and Sarah Dillon (eds), *AI Narratives: A History of Imaginative Thinking About Intelligent Machines* (Oxford: OUP, 2020).
5. Ada Lovelace, 'Translator's Notes to M. Menabrea's Memoir', in Richard Taylor (ed.), *Scientific Memoirs, vol. 3* (London: Richard and John E. Taylor, 1843), 692–731.
6. James W. Cortada, *IBM: The Rise and Fall and Reinvention of a Global Icon* (Cambridge, MA: MIT Press, 2019), 129.
7. Donald Michie, 'Alan Turing's Mind Machines', in Philip Husbands, Owen Holland, and Michael Wheeler (eds), *The Mechanical Mind in History* (Cambridge, MA: The MIT Press, 2008), 61–74.
8. Chris Bernhardt, *Turing's Vision: The Birth of Computer Science* (Cambridge, MA: MIT Press, 2016).
9. B. Jack Copeland (ed.), *The Essential Turing: The Ideas That Gave Birth to the Computer Age* (Oxford: Clarendon Press, 2004).
10. Turing to Ashby (undated), www.alanturing.net/turing_ashby; Turing, 'Intelligent Machinery', in Copeland (ed.), *The Essential Turing*, 431.
11. Stanley Finger, *Origins of Neuroscience: A History of Explorations into Brain Function* (Oxford: OUP, 1994).
12. Flo Conway and Jim Siegelman, *Dark Hero of the Information Age: In Search of Norbert Wiener, The Father of Cybernetics* (New York: Basic Books, 2005).
13. On the deep history of cybernetic thinking, see David W. Bates, *An Artificial History of Natural Intelligence: Thinking with Machines from Descartes to the Digital Age* (Chicago: University of Chicago Press, 2024).
14. Thomas Rid, *Rise of the Machines: The Lost History of Cybernetics* (London: Scribe, 2016), 130–4.
15. Conway and Siegelman, *Dark Hero*, 297–332.

16. Eden Medina, *Cybernetic Revolutionaries: Technology and Politics in Allende's Chile* (Boston, MA: MIT Press, 2014).
17. Rid, *Rise of the Machines*.
18. W. Ross Ashby, *An Introduction to Cybernetics* (London: Chapman & Hall, 1956), 272.
19. 'Lecture on the Automatic Computer Engine', in Copeland (ed.), *The Essential Turing*, 393.
20. 'Computing Machinery and Intelligence', in Copeland (ed.), *The Essential Turing*.
21. Stephen Cave *et al.*, 'The Meanings of AI: A Cross-Cultural Comparison' in Stephen Cave and Kanta Dihal (eds), *Imagining AI: How the World Sees Intelligent Machines* (Oxford: OUP, 2023), 16–37.
22. *The Note-Books of Samuel Butler* (New York: E.P. Dutton & Co., 1917), 42–6.
23. Samuel Buter, *Erewhon* (1872).
24. H.G. Wells, *The War of the Worlds* (London: William Heinemann, 1898), ch. 1.
25. W.R. Ashby, 'Design for a Brain', *Electronic Engineering*, 20:250 (December 1948), 383.
26. Irving John Good, 'Speculations Concerning the First Ultraintelligent Machine', *Advances in Computers*, 6 (1966), 31–88.
27. Cave and Dihal (eds), *Imagining AI*.
28. Vernon Verge, 'Technological Singularity' (1993), in Max More and Natasha Vita-More, *The Transhumanist Reader* (Chichester: Wiley-Blakwell, 2013), 366–75.
29. Joel N. Shurkin, *Broken Genius: The Rise and Fall of William Shockley, Creator of the Electronic Age* (London: Macmillan, 2006), 4.
30. Ibid., 13.
31. Interview with Gordon Moore, 18 August 1995, *Silicon Genesis: Oral Histories of Semiconductor Technology*, Stanford, https://exhibits.stanford.edu/silicongenesis/catalog/hf486bm0643, accessed 12 September 2023.

32. Gordon Moore, 'Solid-State Physicist William Shockley', *Time* magazine (29 March 1999).
33. You can read more about the 'superbaby' sperm bank in Chapter 5 on the history of Mensa.
34. 'On research department productivity and recruiting practices', 18/5/53, ARCH-1995-153, Box 22, folder 17, William Bradford Shockley papers, Stanford University Special Collections.
35. Shurkin, *Broken Genius*, 173.
36. 'Playboy interview with William Shockley, August 1980', in Roger Pearson (ed.), *Shockley on Eugenics and Race: The Application of Science to the Solution of Human Problems* (Washington DC: Scott-Townsend Publishers, 1992), 240.
37. 'Population Control or Eugenics', in Pearson, *Shockley on Eugenics*, 53.
38. Ibid.
39. 'Dysgenics, Geneticity, Raceology: A Challenge to the Intellectual Responsibility of Educators', in Pearson, *Shockley on Eugenics*, 199.
40. Pearson, *Shockley on Eugenics*.

Chapter 8: Augmenting Intelligence

1. Ignacio Ramonet, *Hugo Chávez: my primera vida* (Madrid: Vintage Español, 2014), 420–3.
2. Arnold Hahn, *Use Your Mind: The Road to Successful Thinking* (London: Routledge, 1931).
3. Napoleon Hill, *Think and Grow Rich* (Meriden, CT: The Ralston Society, 1937).
4. Alex F. Osborn, *Applied Imagination: Principles and Procedures of Creative Thinking* (New York: Charles Scribner's Sons, 1953).
5. Jeffrey J. Kripal, *Esalen: America and the Religion of No Religion* (Chicago: University of Chicago Press, 2007), 85.
6. Jane Howard, *Please Touch: A Guided Tour of the Human Potential Movement* (New York: McGraw-Hill, 1970).
7. Glenn Doman, *How to Multiply Your Baby's Intelligence* (New York: Doubleday, 1984), 3.

8. Ibid, 8.
9. Ramonet, *Hugo Chávez*, 420–3.
10. This and other information about de Bono's life and work given below are taken from Piers Dudgeon, *Breaking Out of the Box: The Biography of Edward de Bono* (London: Headline, 2001).
11. Edward de Bono, *The Use of Lateral Thinking* (London: Jonathan Cape, 1967).
12. Luis Alberto Machado, *La revolución de la inteligencia* (Barcelona: Seix Barral, 1975).
13. Luis Alberto Machado, *El derecho de ser inteligente* (Barcelona: Seix Barral, 1978).
14. Office of the Minister of State for the Development of Intelligence, *The Democratization of Intelligence* (1982), box 480, 159.928 A 06 (87), UNESCO Archives (UNESCOA), Paris.
15. Rocío Fernández Ballesteros, Santiago Genovés, Gaston Mialaret, and Hugo Osorio Meléndez, 'Evaluación de los programas de dessarrollo de la inteligencia: Venezuela' (1984), UNESCO Digital Library, https://unesdoc.unesco.org/ark:/48223/pf00000 62176.locale=en (accessed 14 May 2024).
16. Development of Human Intelligence, Part I', 159.928, box 479, UNESCOA
17. Amadou-Mahtar M'Bow to Luis Alberto Machado, 16 June 1979, X07.21 (87) A 8 '79/80', UNESCOA.
18. Dudgeon, *Breaking out of the box*, 216-229.
19. W. Ross Ashby, 'Design for an Intelligence-Amplifier', *Automota Studies*, 34 (1956), 215–34; Peter M. Asaro, 'From Mechanisms of Adaptation to Intelligence Amplifiers: The Philosophy of W. Ross Ashby', in Husbands, Holland and Wheeler (eds), *The Mechanical Mind in History* (Cambridge, MA: The MIT Press, 2008),149–84.
20. Vannevar Bush, 'As We May Think', *Atlantic Monthly* (July 1945), 101–8.

21. Theodor H. Nelson, 'As We Will Think', in James M. Nyce and Paul Kahn (eds), *From Memex to Hypertex: Vannevar Bush and the Mind's Machine* (San Diego, CA: Academic Press, 1991), 245–60.
22. James M. Nyce and Paul Kahn, 'A Machine for the Mind: Vannevar Bush's Memex', in James M. Nyce and Paul Kahn (eds), *From Memex to Hypertex: Vannevar Bush and the Mind's Machine* (San Diego, CA: Academic Press, 1991), 39–67.
23. Nyce and Kahn, *From Memex*, 235–6.
24. Thierry Bardini, *Bootstrapping: Douglas Engelbart, Coevolution, and the Origins of Personal Computing* (Stanford, CA: Stanford University Press, 2000), 13.
25. D.C. Engelbart, 'Study for the Development of Human Intelligence Augmentation Techniques', February 1967, series 2, accession 1998-094, 1967, Douglas Engelbart papers (DEP), SUSC.
26. 'Augmented Human Intellect Program', 1 March 1962, series 1, accession 1986-030, box 5, folder 1, DEP, SUSC.
27. Engelbart, 'Study for the Development of Human Intelligence Augmentation Techniques'.
28. Bardini, *Bootstrapping*, 200–2
29. *Ibid.* 194.
30. Bush, 'As We May Think', 104.
31. Stephanie K. Bell, Jayce C. Lucke and Wayne D. Hall, 'Lessons for Enhancement from the History of Cocaine and Amphetamine Use', *AJOB Neuroscience*, 3:2 (2012), 24–9.
32. Nicolas Rasmussen, *On Speed: The Many Lives of Amphetamine* (New York: New York University Press, 2008), 28–31.
33. Ibid., 64–85.
34. *Boosting your brainpower: ethical aspects of cognitive enhancements. A discussion paper from the British Medical Association* (London: British Medical Association, 2007).
35. Ibid., 18.

36. *Human enhancement and the future of work: Report from a joint workshop hosted by the Academy of Medican Sciences, the British Academy, the Royal Academy of Engineering and the Royal Society* (The Academy of Medical Sciences, October 2012).
37. Max More and Natasha Vita-More (eds), *The Transhumanist Reader: Classical and Contemporary Essays on the Science, Technology, and Philosophy of the Human Future* (Oxford: Wiley-Blackwell, 2013).
38. Alison Bashford, 'Julian Huxley's Transhumanism', in Marius Turda (ed.), *Crafting Humans: From Genesis to Eugenics and Beyond* (Goettingen: V&R unipress, 2013), 153–68.
39. Julian Huxley, *UNESCO: Its Purpose and Philosophy. Facsimiles of English and French editions of this visionary policy document* (London: Euston Grove Press, 2010), 18–21.

Chapter 9: Resisting and Rethinking Intelligence

1. 'Haringey Comprehensive Schools', 13 January 1969, Black Education Movement (Early Period, 1965–1988) – GB 2904, BEM/1, George Padmore Institute Archives (GPIA), Finsbury Park, London.
2. 'Report to the Education Committee on Comprehensive Education', March 1969, BEM 1, GPIA.
3. NLWIA to Haringey Education Committee, December 1969, BEM 1/1/2, file 2, GPIA.
4. Report on Working Luncheon with Alderman Doulton as Vice-Chairman of the Haringey Education Committee at West Indian Club in Whitehall, 13 May 1969, BEM 1/1/1, GPIA.
5. 'Black parents fight for your rights', *Black Dimension*, 1:3 (1969), 13–14, GPIA.
6. 'War against black children', BEM 1/2/4, GPIA.
7. BEM 1/2/6, GPIA.

8. Simone de Beauvoir, *The Second Sex* (London: Jonathan Cape, 1953), 658.
9. Tom Harrisson, founder of Mass Observation, quoted in Selina Todd, *Snakes and Ladders: The Great British Social Mobility Myth* (London: Chatto & Windus, 2021), 125.
10. Mandler, *The Crisis of Meritocracy*.
11. Brian Simon, *Education and Social Order, 1940–1990* (New York: St Martin's Press, 1991); Brian Simon, *A Life in Education* (London: Lawrence & Wishart, 1998).
12. Michael Young, *The Rise of the Meritocracy* (New Brunswick, NJ: Transaction, 1994).
13. Douglas M. McIntosh, *Educational Guidance and the Pool of Ability* (London: University of London Press, 1959)
14. Jill Pellew and Miles Taylor (eds), *Utopian Universities: A Global History of the New Campuses of the 1960s* (London: Bloomsbury, 2021).
15. A.H. Halsey (ed.), *Aptitude Intellectuelle et Education* (Paris: OCDE Publications, 1962).
16. OECD, *Towards Mass Higher Education: Issues and Dilemmas* (OECD: Paris, 1974).
17. Christopher Newfield, 'California Dreaming: Clark Kerr and the University of California', in Pellew and Taylor (eds), *Utopian Universities*, 251–68.
18. Eric Ashby, *Any Person, Any Study: An Essay on Higher Education in the United States* (New York: McGraw Hill, 1971).
19. Kingsley Amis, 'Lone Voices', *Encounter* (July 1960), 8.
20. Ashby, *Any Person*, 91.
21. OECD, *Towards Mass Higher Education*, 88.
22. Richard Lynn, 'Comprehensives and Equality: The Quest for the Unattainable', *Critical Survey*, 4:3 (1969), 26–33.
23. Hazel V. Carby, 'Schooling in Babylon' in Centre for Contemporary Studies (eds), *The Empire Strikes Back: Race and Racism in 70s Britain* (Hutchinson: London, 1982), 184.

24. R.J. Holdman and Francine M. Taylor, 'Coloured immigrant children: a survey of research, studies and literature on their educational problems and potential – Britain', *Educational Research*, 8:3 (1966), 163–83; Philip E. Vernon, 'Environmental handicaps and intellectual development: part I', *British Journal of Educational Psychology*, 35:1 (1965), 9–20.
25. Bernard Coard, *How the West Indian Child is Made Educationally Subnormal in the British School System: The Scandal of the Black Child in Schools in Britain* (London: New Beacon Books, 1971).
26. Ibid., 19.
27. On the early Black education movements, including the complex history of supplementary schools, see Paul Warmington, *Black British Intellectuals and Education: Multiculturalism's Hidden History* (London: Routledge, 2014), chs. 3 & 4.
28. Michael E. Staub, *The Mismeasure of Minds: Debating Race and Intelligence between Brown and The Bell Curve* (Chapel Hill: University of North Caroline Press, 2018), 14–16.
29. Arthur Jensen, 'How much can we boost IQ and scholastic achievement?', *Harvard Educational Review*, 39:1 (1969), 1–123.
30. Shurkin, *Broken Genius*.
31. 'King's College London enquiry into publications authored by Professor Hans Eysenck with Professor Ronald Grossarth Maticek', May 2019, https://www.kcl.ac.uk/assets/news-statements/hans-eysenck-enquiry-final-may-2019.pdf (accessed 30 April 2025).
32. Hans Eysenck, *Race, Intelligence and Education* (London: Temple Smith, 1971).
33. Roderick D. Buchanan, *Playing With Fire: The Controversial Career of Hans J. Eysenck* (Oxford: OUP, 2010), 273.
34. Richard Herrnstein, 'I.Q.', *The Atlantic* (September 1971), 44–64. Herrnstein later expanded these ideas in *I.Q. in the Meritocracy* (Boston: Atlantic Monthly Press, 1973).

35. Buchanan, *Playing with Fire,* 298–302.
36. Quoted in H.J. Eysenck, 'The Dangers of the New Zealots', *Encounter* (December 1972), 79–90.
37. N.J. Block and Gerald Dworking, *The IQ Controversy: Critical Readings* (New York: Pantheon Books, 1976).
38. Bernard Coard, 'Race, intelligence and education', *The Times Educational Supplement,* 2926 (18 June 1971), 4.
39. Richard J. Herrnstein and Charles Murray, *The Bell Curve: Intelligence and Class Structure in American Life* (New York: The Free Press, 1994).
40. Staub, *The Mismeasure,* 156–7.
41. Ibid., 160–72.
42. Kahneman discusses the origins of his theories and his work with Tversky in the introduction to Daniel Kahneman, *Thinking, Fast and Slow* (New York: Farrar, Straus and Giroux, 2013).
43. Howard Garnder, *Frames of Mind: The Theory of Multiple Intelligences* (New York: Basic Books, 2011 [first published 1983]).
44. Gardner, *Frames of Mind,* 10.
45. Ibid., 4.
46. Ibid., xvii–xviii.
47. Daniel Goleman, *Emotional Intelligence: Why It Can Matter More than IQ,* 25th anniversary edition (London: Bloomsbury, 2020).
48. Ibid., 78–81.
49. (The aptly named) Paul Tough, *How Children Succeed: Grit, Curiosity, and the Hidden Power of Character* (2012).
50. Staub, *The Mismeasure,* ch. 4.
51. On the history of autism and neurodiversity, see Steve Silberman, *NeuroTribes: The Legacy of Autism and the Future of Neurodiversity* (New York: Avery, 2015).
52. Leo Kanner, 'Autistic disturbances of affective contact', *Nervous Child,* 2 (1943), 217–50.

53. Harvey Blume, 'Neurodiversity: On the neurological underpinnings of geekdom', *The Atlantic* (September 1998).
54. Kanner, 'Autistic disturbances', 248.
55. Hans Asperger, '"Autistic psychopathy" in childhood', in Uta Frith (ed.), *Autism and Asperger Syndrome* (Cambridge: Cambridge University Press, 1991), 37–92.
56. Kanner, 'Autistic disturbances', 247.
57. For examples of such criticisms, see Mohen Costandi, 'Against neurodiversity', *Aeon* (12 September 2019); Bonnie Evans, 'After neurodiversity', *Aeon* (29 July 2021).

Chapter 10: Why the History of Intelligence Matters
1. John Maynard Keynes, *The General Theory of Employment Interest and Money* (London: Macmillan, 1936), 383.
2. Stephen Cave, 'The Problem with Intelligence: Its Value-Laden History and the Future of AI', *Proceedings of the AAAI/ACM Conference on AI, Ethics, and Society* (2020), 29–35.
3. Tony Blair, *On Leadership: Lessons for the 21st Century* (New York: Crown, 2024), 135.

Index

2001: A Space Odyssey (S. Kubrick & A.C. Clarke) 253–4

A
Abbasid Caliphate (750–1258) 69
abolitionists 89
Adam Smith Institute (ASI) 183–5
African-American education 121–4, 151–2
African education system, colonial 150–1, 152–4
age-specific mental ability 103–4
Akademgorodok 207
Al-Farabi 70–1
Allen, Paul 29
Allende, Salvador 241
Alles, Gordon 299–300
Alpha Brain 3
Amazon 33
American Psychological Association 133–4, 135–6, 275
Amis, Kingsley 322
amphetamines 300–2
Analytical Engine 229–31
anti-colonialism 149–50, 153–4
anti-Semitism 156
Apple 29, 296
Aristotle 66, 68
Army, British 137, 301
Army intelligence tests, US 111, 133–6, 148
ARPANET 295
artificial general intelligence (AGI) 362
artificial intelligence (AI) 3–4, 6, 58, 60, 226–9, 242–5
   and beliefs in human progress 361–6
   fear of machine superintelligence 246–8, 250–6, 268

Ashby, Ross 225–7, 239, 242, 243, 250, 287–8
Asimov, Isaac 161, 254
Asperger, Hans 353, 354
'Asperger's syndrome' 354
Atlantic magazine 289, 335, 338
Atlas Fellowship 58
Attlee, Clement 23
augmenting human intelligence 269–71, 305–6, 363
   Edward de Bono the and promotion of thinking skills 278–81, 283, 284, 286
   Glenn Doman and the Better Baby Institute 275–7
   Human Potential Movement 273–4
   mental hygiene 271–2
   smart drugs and cognitive enhancement 298–302
   through computer/digital technology 286–98
   transhumanism 303–4
   Venezuelan Ministry for the Development of Intelligence 269–70, 281–5, 343, 363
autism and autism spectrum disorder 352–7
Avicenna 70

B
Babbage, Charles 229–30
babies/infants, intelligence in 5, 275–6, 282, 340
Bacon, Roger 229

banking/financial sector 26–7
Bankman-Fried, Sam 57–9
Bardeen, John 260
BBC 146, 163, 336
Beer, Stafford 241
Bell Curve (R. Hernnstein & C. Murray) 18, 53, 54–5, 62, 339–41, 346
Bell, Daniel 20–1, 22
Bell Laboratories 260, 263
Bennett-Goleman, Tara 346–
Bernard van Leer Foundation 343
Berrill, Roland 163–5, 170–1, 179
Better Baby Institute 276
Bezos, Jeff 3
Biden, Joe 2
billionaires, cult of super-intelligent 48–61
Binet, Alfred 100–6, 107, 132
Binet-Simon tests 103–6, 107, 109, 112
birthrates 55–6, 100, 181, 265
Black Dimension 311
Black education movement, UK 311, 326–9
Black elitism 121–4
Black Power movement 266, 329
'Black uplift' 123–4
Blair, Tony 4, 22, 36, 361, 364
Bletchley Park 232, 233, 235
Blouin, Andrée 222–3
Board of Education, British 145–6
Bolshevik revolution 196–8
Boole, George 238
brain function, human 194–6, 228, 237–8, 352–8, 364
'Brain Trust,' US government 22–3
Branch, Margaret 211–12
Brattain, Walter 260
Brave New World (A. Huxley) 252, 273
breeding programmes, eugenics and controlled 155–7, 267
breeding, selective 56–7, 68, 181–2, 263
Brown, Gordon 23
Brown vs Board of Education ruling, UK 330
Buchanan, Pat 53
Burt, Cyril 138, 146, 163, 174–5, 323
Bush, George H.W. 54

Bush, Vannevar 288–91
business literature, thinking skills and 272–3, 278–9, 280
Butler, Eamonn 184
Butler, Samuel 246–7

C
C9 League universities, Chinese 39
Cadbury, George 145
CAF (Consider All Factors) 'thinking skills' tool 278, 281
Calogero, Don 187–93, 343
Campins, Luis Herrera 282
Caribbean Education and Community Workers Association (CECWA) 329
Carlyle, Thomas 81
Carter, Cedric 180–1
Cattell, James 42–3
Cave, Stephen 363
Centre for Effective Altruism 57–8
ChatGPT-4 3–4
Chávez, Hugo 269–70, 278, 284
Chicago, University of 39
children
  Alfred Binet's tests 100–6
  American spelling bees 220–2
  babies and infants 308–11, 313–19, 323–34, 340
  Don Calogero's school for gifted children 188–93, 343
  gender biases in education 217–18
  gifted children in communist society 193–210, 216–17
  gifted children movement in the west 173–8, 210–19, 340, 343
  Glenn Doman and the Better Baby Institute 275–7
  identification of autism 352–3, 357
  impulse control tests and EQ 347–8, 350
  intelligence and racial biases 218–19, 223–4, 308–11, 313–19, 323–34, 340
  Lewis Terman's tests 108–11
  Mensa on the education of gifted children 173–8

school education 5, 102–3, 141–54, 173–8, 188–93, 196–219, 223–4, 281, 283, 286, 308–11, 313–19, 323–34, 340, 341, 345–6, 366
vocational guidance 138–9
see also racism
China 345
  examinations and intellectual meritocracy 72–7, 78–9, 143
  gaokao system 39, 78
  university education 38, 39
Christianity 79–80, 148, 151
Churchill, Winston 301
Civil Rights movement 266, 321, 329, 330
Civil Service, British 143
Clarke, Arthur C. 254
class conflict, post-WWI 139–40
Clinton, Bill 23, 36
Clinton, Hillary 340
Coard, Bernard 326–9, 338
cocaine 299
cognitive elites 15–20, 339, 359
  elite education 35–47
  knowledge economy 20–34
  neoliberal supermen 48–61
cognitive enhancement, smart drugs and 298–302
Cold War 23, 43, 216–17
collective intelligence 352
colonialism 85–6, 115, 147–54, 249
Colossus computers 235–6
communist society, education and gifted children in 193–210
comprehensive schools, rise of 316–19, 323, 334, 341
computer technology
  Alan Turing 232–4, 236, 237, 242, 243–5, 250
  augmenting human intelligence 286–98
  Colossus computers 235–6
  cybernetics 239–42
  development of transistors 256, 258–9, 260
  early development of 227–8, 229–32
  and human brain function 237–9, 268
  John von Neumann 236
  Second World War 234–6, 240, 288
  Tommy Flowers 235
  see also artificial intelligence (AI); Silicon Valley; tech sector
Coolidge, Calvin 221
CoRT (Cognitive Research Trust) 281, 283
counter-culture, Californian 30, 273, 296
Covid-19 pandemic 24, 366
Cox, Catherine 112
Crane, Jennifer 215
craniology 9
creativity and success 352
crowds, psychology of 125–8
cryptocurrency 57
Cummings, Dominic 2
curiosity quotient (CQ) 352
cybernetics 239–42

D
Daily Mail 226
Danna, Jo 191
Darwin, Charles 88, 95, 247
de Beauvoir, Simone 312
de Bono, Edward 269, 270, 278–81, 283, 284, 286
de Gaulle, Charles 126
de Rivera, General Miguel Primo 128–9
DEI programmes 60
Department of Government Efficiency, US government 49
Diderot 81, 82
diet and intelligence 5
Difference Engine 229–30
'The Disaster', Spain 128
divine intelligence 64, 71, 91
Doman, Glenn 275
Douglass, Frederick 89–90
Doulton, Alfred 307–11, 326
Draghi, Mario 24
drugs 30, 273, 298–302

Du Bois, W.E.B. 121–4, 130, 151–2
Durov, Pavel 56
dysgenics 180–1, 339–40

E

Economist, The rearrange 17, 27
'educational ladder' 144
educational reform, egalitarian 313–24, 341
Educational Testing Service (ETS) 142
Edwards, Peter 192
effective altruism 57–9
effort and work ethic 46–7
egalitarianism 182–3, 278, 312–13, 339, 341
elite education
  elite universities 38–41
  increase of university education 37–8
  as political imperative 35–7
  university rankings 41–6
  see also school education
Ellis, Havelock 114
emotional intelligence (EQ) 18–19, 31, 342, 346–52
'Empower Exceptional People' funding stream, FTX 58
Engelbart, Douglas 291–3, 295–7
Engels, Friedrich 197
ENIAC computer 236
Enlightenment, the rearrange 79–92, 113, 115
Enron 15–17, 33–4
entrepreneurship 50–3, 55
Entscheidungsproblem 233–4
Equality of the Human Races (A. Firmin) 90
Equiano, Olaudah 89
Erewhon (S. Butler) 247–8
Esalen 274–5
ESN schools 326–9, 338
Essay on the Inequality of Human Races (A. de Gobineau) 87–8
eugenics 9, 42–3, 56, 154–5, 218, 304–5
  in Africa 150–1
  challenges against 157–8
  enforced sterilisation 154–7
  Francis Galton 95–100
  Lewis Terman 111
  Mensa associations 180–2
  William Shockley 259, 264, 265–8
Eugenics Review 334
Eugenics Society 111, 150, 180
European Union (EU) 24
Eurozone crisis (2011) 24
euthanasia 157, 182
evolution 88, 247
extreme work, cult of 46
Eysenck, Hans 323, 333–4, 336–8, 340

F

Fabian Society 145, 182, 319
Fairchild Semiconductor 262
far-right beliefs 53–5, 179
  see also Nazi Germany; racism
feminism 90–1, 312
feudal elites, European 77
Firmin, Anténor 90
First World War 111, 131, 133, 137
Flowers, Tommy 235
Founders Fund 29
founders, tech 29–31
Franco, Franciso 128
free-market-economics 183–4, 185
Freud, Sigmund 299
FTX and the Foundation Future Fund 57–9
Fuller, Buckminster 161, 284
Fuller, Tom 89

G

Galen 69, 237
Galton, Francis 48, 95–100, 113, 155
Gardner, Howard 307, 342–4
Gates, Bill 29, 31, 36
Gauss, Carl Friedrich 195
gender biases 90–1, 112–14, 115, 217–18, 312
General Electric 252
'general intelligence' (*g*) 107–8, 145–6
genetics 53–4, 200, 264, 270, 304, 340–1
  see also eugenics; intelligence science; racism
'genius,' use of the term 80–1
Germany, East 209–10

G.I. Bill, US 319
Gladwell, Malcolm 34
globalisation 35–7
Gobineau, Arthur de 87–8, 90, 92
Goddard, Henry 116, 132–3, 140
'Golden Age,' Islamic 69–71
Golden, Marie 220
Golem 229, 246
Goleman, Daniel 346–7, 348
Good, Irving John 250–1, 254
Google DeepMind 4
Gorky, Maxim 196
Gould, Stephen J. 9
Graham, Robert 181–2
Greece, ancient 66–9, 70
Grotius, Hugo 86

H
Hadid, Bella 3
Hague, William 26
Haringey protests (1969) 307–11, 324, 326, 329, 336
Harris, Kamala 2–3
Harvard Business School 26, 27
Harvard Educational Review 331
Harvard University 39, 40, 283, 335, 343
Hassabis, Demis 4
Hawking, Stephen 161
Helvétius 82–3
Hereditary Genius (F. Galton) 96, 99
hereditary intelligence, belief in 96–100, 109–10, 123–4, 270–1, 278, 335
Herrnstein, Richard 62, 283, 334–7, 338–40
higher education
see university education
Hitler, Adolf 126, 157, 300
Hollingworth, Leta Stetter 114–15, 187, 211–12, 214–15, 216, 218
homeostat machine 225–7, 239, 243, 287
'homunculus,' creation of 229
Hopkins, John 352
Hugo, Victor 81
Human Genome Project 54
Human Potential Movement 273–4
Huxley, Aldous 252, 269, 273–4, 299
Huxley, Julian 180, 273, 304–5

I
IBM 137, 232, 236
Ibn Khaldun 70
'idiots' and 'imbeciles,' categorisation of 85–6, 104, 356
Iliad (Homer) 228–9
immigration/immigrants 155–6, 308–9, 324–5
impulse control 347–51
India, university education 38, 39–40
Indian Civil Service 143, 148
Indian school education, colonial 147–9, 153
industrial psychology 137–9
Industrial Revolution 124–5, 185
inequality and intelligence 11–13, 17–18, 54–5, 83–92, 121–4, 115, 175–6, 183, 185, 306, 339–40, 349–50, 365, 366–7
see also racism
Intel 262
'intelligence boosting' products 3
intelligence, rethinking 341–2
  Daniel Goleman and EQ 346–52
  Harold Gardner and MI theory 342–6, 351–2
  neurodiversity 352–8
intelligence science 9, 11, 18–19, 27–8, 42–3, 55, 65, 93–5
  Alfred Binet 100–6, 107, 132
  Charles Spearman 107–8, 145
  criticism of 157–8, 311–13, 337–8
  Cyril Burt 146–7, 323
  Francis Galton 95–100, 113, 155
  gender biases 90–1, 112–14, 115, 217–18, 312
  Henry Goddard 132–3
  Lewis Terman 108–12, 113, 134–5, 148–9, 155, 157–8, 211, 213–15, 217–18, 257, 259, 312
  rethinking intelligence 278, 317–23, 341–58
  W.E.B. Du Bois 121–4
  workplace recruitment 116–21, 132–40
  see also eugenics; IQ (intelligence quotient); racism; school education

International Health Exhibition (1884) 96
International Mathematical Olympiad (1959), Romania 209
IQ (intelligence quotient) 1, 2, 4–5, 9, 18–19, 25, 31, 53–4, 55, 108–12, 113–14, 119, 124, 147, 162, 166, 173–4, 184, 202, 213, 259, 265–6, 310–11, 327–8, 331–3, 335, 341, 342, 344, 349, 364
see also Mensa
Islamic philosophers, early 66, 69–71
Italy 23–4
Ivy League institutions 41

J
Japanese hierarchical society 77
Jefferson, Thomas 83
Jensen, Arthur 331–4, 336–7, 340
job losses, AI and potential 365–6
Jobs, Steve 28–30
Johnson, Boris 1–2
Johnson, Lyndon 321, 330
Johnson, Samuel 90
Jones, Thomas Jesse 151–2

K
Kahneman, Daniel 341
Kanner, Leo 352–3, 357
Kansas State University 322
Kant, Immanuel 81
Kennedy, John F. 23
Kenya 150
Kenyatta, Jomo 154
Kerouac, Jack 301
Keynes, John Maynard 360–1
Khrushchev, Nikita 206–7
Kite, Elizabeth 112
knowledge economy
  banking industry 26–7
  Enron 15–17, 33–4
  McKinsey and management consultants 24–6, 31–3
  politics 22–4, 40–1
  post-industrial society 20–2
  talent management 31–2
  tech industry 27–31

Krupskaya, Nadezhda 197–8
Kubrick, Stanley 253–4

L
Labour Party, UK 23, 37, 318, 334
lateral thinking 279–80
Latin American Centre for the Development of Intelligence 285
Lavrentyev, Mikhail 207–8
Le Bon, Gustave 125–8, 129
learning disabilities 84–5, 102–5, 157, 326–7
Leibniz, Gottfried Wilhelm 229
Lenin, Vladimir 194–6, 197
Linnaeus, Carl 84
Lippmann, Walter 157–8
Locke, John 84
Loftus, Jamie 179
Lovelace, Ada 230–1
Luria, Alexander 200–2
Lynn, Richard 323

M
Machado, Luis Alberto 281–5, 343
Magnus, Albertus 229
Malaysia, University of 45
male bias, intelligence science and 113–14
management consultation 24–6
management theories, scientific 50–1
Manchester Baby and Manchester Mark 1 236
Mankind Quarterly 177–8
March, James 41
Margolin, Leslie 218–19
Markovits, Daniel 46
'marshmallow test,' delayed gratification and the 347–8, 350
Marxism 197
Maslow, Abraham 275
Mass Observation 119
'masses,' concern about the 124–31
Maudsley Institute of Psychiatry 172
McCarthy, John 245
McKinsey 25–6, 31–3, 34
Mead, Margaret 191, 240
MEI (merit, excellence and intelligence) 60

Memex machine 290–1
Mensa 25, 59, 160–2, 185–6, 363
  early years 163–6
  on the education of gifted
    children 173–8, 192–3
  higher aspirations 170–8
  joining and membership 166–70
  politics and eugenics 178–85
'mental engineering' 121, 136
'mental hygiene' 271–2
meritocracy 28, 46–7, 60, 72–9, 143, 183, 318–19
metis 67
Microsoft 29
Mill, John Stuart 93
Minsky, Marvin 304
Mischel, Walter 347–8, 350
missionaries, Christian 148, 151
Mockerie, Parmenas 154
Mont Pelerin Society 51
Monti, Mario 24
Moore, Gordon 262
'morons' 104
multiple intelligences (MI) 18–19, 31, 342–5, 351–2
Münsterberg, Hugo 136
Murray, Charles 54, 62, 338–40
Musk, Elon 3, 49, 55–6, 60, 361
Mussolini, Benito 126, 128
Myers, Charles 137–8

N
NASA (National Aeronautics and Space Administration) 293, 295
National Defence Education Act (1958) 216–17
National Front 307
National Institute for Industrial Psychology (NIIP) 137–9
Nazi Germany 156–7, 300
Nego Problem, The rearrange (W.E.B. Du Bois) 122–4
Nehru, Jawaharlal 241
Nelson, Ted 291
Neo-Confucianism 74
neoliberal supermen 48–61
neoliberalism 35, 53, 162, 183

neurodiversity 18–19, 342, 352–8
neuroscience 54, 228, 237–9, 273–4, 352
New Deal 22–3
New York Times 300, 347
Nixon, Richard 331
Nobel Prize for Physics 259, 260
nous 66
Noyce, Robert 262

O
Obama administration 40–1
Obama, Barack 2, 23, 47
OECD (Organisation for Economic Co-operation and Development) 6
Olympiad system, scientific 205–9, 216
online communities, autistic 354–
Open University 321
Ortega y Gasset, José 129–30
Osborn, Alex 272–3
Oxford University 4

P
Palo Alto, California 256–8
Pascal, Blaise 229
Pavlov, Ivan 196
pedology ban, Soviet 203–5
personalities, evaluation of 101–2
Peterson, Jordan 55
Phelps-Stokes Fund 151–2
philanthropy, effective altruism and 57–9
philosophers, ancient 66–71
physical disabilities 157
Pirie, Madsen 183–5
Plato 66, 67–8
Playboy magazine 264
Player Piano (K. Vonnegut) 252–3
PMI (Plus, Minus, Interesting) 'thinking skills' tool 278, 281
politics and knowledge economy 22–4, 40–1
post-industrial society 20–2
Powell, Enoch 338
pre-19th-century understanding of intelligence 62–5
  Chinese meritocracy 71–9
  reason and the philosophers 66–71

405

pre-Adamites 86
'Project on Human Potential,' Harvard 343
'Project Zero,' Harvard 343
Protestants 80, 82
psychedelic drugs 273, 299
Ptolemy 69

Q
Qing era, Chinese 72
quick thinking, Renaissance value of 82
Qur'an 70

R
racism 3, 12, 115
   colonial 85–6, 150–1, 152–4, 249
   gifted children movement 218–19
   Haringey protests (1969) 307–11, 324, 326, 329, 336
   Nazi Germany 156–7
   scientific and educational 9, 28, 86–90, 92, 110–11, 150–1, 152–4, 177–8, 259, 264, 265–8, 307–11, 313, 324–40, 346, 351
Rain Man film (1988) 354, 357
Ramón y Cajal, Santiago 238
rankings, university 41–6
Reagan, Ronald 217
reason
   and early philosophy 66–71
   and religion 69–71, 79–80
Reformation 80
reformists, social 144–5
Reich, Robert 21
Renaissance 82
Repository for Germinal Choice 181–2
resilience and 'grit' 349
Richey, James Alexander 148–9
Rickover, Admiral Hyman G. 216
Robbins Report (1963) 321
Rogan, Joe 3, 49
Roosevelt, Franklin D. 22–3
Roosevelt, Theodore 126
Russian Revolution 139

S
salaries, intelligence and 6, 32, 46, 366
Salpêtrière hospital, Paris 102
Sandel, Michael 23
Saudi Arabia 39
savants, autistic 356–8
Scholastic Aptitude Test (SAT), US 142
school education 5, 102–3, 141–54, 173–8, 187–93, 196–219, 281, 283, 286, 308–11, 313–19, 323–34, 340, 341, 345–6, 366
   see also racism; university education
Schumpeter, Joseph 51
science fiction 248–9, 251–6
scientific management 132
Second World War 232, 234–5, 240, 242, 259–60, 288, 300–1, 329
self-control/delayed gratification 347–51
'self-help' literature, mental 272
Serebriakoff, Victor 165, 168, 171, 173–4, 175–7, 181, 182, 185, 192
Shanghai Jiao Tong University 44
Shannon, Claude 238
Shockley Semiconductor Laboratory 258–9, 261–2, 263
Shockley, William 28, 181, 258–68, 332–3
Silicon Valley 27–31, 256–8, 262–3, 356
Simon, Brian 317
Simon, Théodore 103
Simpsons, The rearrange 160–1, 186
Sinclair, Clive 167, 183, 185
'singularity' 255
Six Thinking Hats (E. de Bono) 280
Skilling, Jeffrey 16–17
Skinner, B.F. 284
slavery 86, 89
Slobodian, Quinn 53
smart drugs and cognitive enhancement 298–302
'smart,' use of the term 19
Social Dawinism 88
Society for the Psychological Study of the Child 102
socio-economic inequality
   see hereditary intelligence, belief in; inequality and intelligence; racism

Song dynasty, Chinese 75
Soviet Union 194–210, 241
Spanish Civil War 129, 317
Spanish colonies 128
Spanish Republic 128–9
Spearman, Charles 107–8, 138, 145
spelling bees, American 220–2
Spencer, Herbert 88
sperm banks 56–7, 181–2, 263, 363
spirituality, reason and 67, 91
Sputnik launch, Soviet 206
Stalin, Joseph 204–5, 206
Stanford-Binet tests 110–12, 134, 138, 149
Stanford Research Institute 293, 295–6
Stanford University 28, 38, 108, 257, 348
Staub, Michael 349
sterilisation procedures 155–7
Stern, William 109
Students for a Democratic Society 336
suffrage 126
Summers, Larry 115
superintelligence, fear of machine 246–8, 250–6, 268
syllogisms 200–1
symbolic analysts 21–2

T
T4 euthanasia programme, Nazi 157
'talent,' recruitment of 31–4, 40
talented tenth, the 121–4, 151–2, 193
Taylor, Frederick 24, 50–1, 131
tech sector 27–31
  see also artificial intelligence (AI); computer technology; Silicon Valley
technocracy 22–3
Terman, Lewis 107, 108–12, 134–5, 148–9, 157–8, 211, 213–15, 217–18, 259, 312
test scores, negative impact of low intelligence 116–20, 158–9, 344
testing intelligence
  see eugenics; intelligence science; IQ (intelligence quotient), school education
Thatcher, Margaret 185

The Times Higher Educational Supplement 44, 45
Thiel, Peter 29
thinking skills, Edward de Bono and 278–81, 283, 284, 286
Thompson, Helen Bradford 114–15
Thorndike, Edward 155
Time magazine 226, 259
transhumanism 303–4
tripartite educational model, British 314–16
Trotsky, Leon 196
Trump, Donald 2–3, 60
Turing, Alan 232–4, 236, 242, 243–5, 248, 250
'Turing test' 244–5
Tusk, Donald 55
Tuskegee Institute, Alabama 152, 154
Tverskey, Amos 341

U
UNESCO 180, 285
United Kingdom
  Black education movement 307–11, 326–9, 336, 338
  colonial Africa 150–1, 152–4
  colonial India 143, 147–9, 153
  education of gifted children 211–12, 215
  intelligence testing and workplace recruitment 137–9
  Labour government 23, 37, 318, 334
  Mensa 163–5, 173–7, 180–1, 183–5
  school system 143–54, 175–6, 314–19, 323, 324–9, 334, 341
  university education 23, 37, 321, 322
United States of America 220
  African-American education 121–4, 151–2, 154
  Army intelligence tests 111, 133–6, 148
  Binet-Simon tests 104–5, 107
  education of gifted children 210–11, 216–17

United States of America (*cont.*)
  elite education 38, 40–3
  enforced sterilisation 155–6
  Enron 15–17, 33–4
  intelligence testing and workplace recruitment 131–7
  McKinsey 24–6, 31–3, 40
  Mensa 165, 173, 177–8, 182, 183
  National Defence Education Act (1958) 216–17
  national scientific competitions 209
  politics and knowledge economy 22–3
  school education system 121–4, 142, 147, 151–2, 154, 210–11, 216–17, 330–3, 334, 340, 345–6
  Silicon Valley 27–31, 256–8, 262–3, 356
  university education 38, 40–3, 319–20, 321–2
University College London 99, 107, 172
university education 6, 23, 35–8, 366
  elite education 38–47
  increased access to 319–23
  protests against racism 336–7
university rankings 41–6
U.S. News & World Report 44–5, 266
Uzbekistan, Soviet 200–1

V
Valéry, Paul 130–1
'variability' theory and male intelligence 114–15
Venezuelan Ministry for the Development of Intelligence 269–70, 278, 281–5, 343
venture capitalists 29
Villaggio del Superdotato, Sicily 188–93, 343, 363
Vinge, Vernor 255

vocational guidance 138–9
Vogt, Oskar 194
Voltaire 82–3
von Neumann, John 236
Vonnegut, Kurt 252–3

W
Wang, Alexandr 60
War of the Worlds (H.G. Wells) 248–9
Ware, Lancelot 163, 164
Washington, Booker T. 123, 152
'Way Learning,' Chinese 74, 75
Webb, Beatric 145
Wells, H.G. 249
Weyl, Nathaniel 177–8, 180, 182
white supremacists 3
  see also Nazi Germany; racism
Wiener, Norbert 239–41, 242, 253
Wing, Lorna 353–4
Wollstonecraft, Mary 90–1
women's intellect, beliefs in 90–1, 112–14, 115, 217–18
Wordsworth, Elizabeth 160, 162
work, cult of extreme 46
workplace, social engineering in the 132–40
Wundt, Wilhelm 94

X
Xerox Palo Alto Research Center (PARC) 296

Y
Y Combinator 29
Yale University 41
Yerkes, Robert 135–6
Young, Michael 318–19, 335

Z
Zetkin, Clara 196
Zuckerberg, Mark 28